Praise for
ALIEN Thinking

'A sharp critique of the conventional wisdom around innovation – with thought-provoking advice for how to do it better' Jake Knapp, inventor of the Design Sprint and *New York Times* bestselling author

'Apply *ALIEN Thinking* to go from early opportunity recognition to impactful innovation. I really enjoyed how this book offers compelling insights and powerful illustrations to enhance the innovation process' Alex Osterwalder, author of the two-million-copy bestseller *Business Model Generation*

'Stuck in innovation stagnation? Read this brilliant book and you'll break free from it' Rolf Dobelli, author of the million-copy bestseller *The Art of Thinking Clearly*

'*ALIEN Thinking* is an accessible approach to finding and implementing breakthrough ideas. The authors offer illuminating examples to introduce and illustrate the real-world efficacy of each of the major concepts. I recommend this book to anyone willing to do the hard work of generating and implementing breakthrough ideas' Bernard Roth, academic director of the Stanford d.school and author of *The Achievement Habit*

'Compared to many innovation books, this one is unique in that it emphasizes the emotional side of innovation. Practising innovators will find a sympathetic companion in this fascinating book' Keith Sawyer, author of *Zig Zag* and *Group Genius*

'In today's fast-changing world, innovation is essential for success. The authors of *ALIEN Thinking* have drawn upon their extensive experience to provide a powerful new framework to dramatically improve the odds for innovative thinking and breakthroughs' Robert Rosenberg, retired CEO of Dunkin' Donuts and author of *Around the Corner to Around* '' *World*

'I wish I had read this book years ago. A very pragmatic and e framework to help make innovation a game-changing reality, are in a small start-up or a big multi-national corporation' Cl executive vice president of Nestlé, chief executive officer Oceania and sub-Saharan Africa

'*ALIEN Thinking* offers a business-oriented, applicabl innovate. It is a must-read for corporate innovators and create true value for their customers and organization. Nussbaumer, head of corporate innovation, SIGVARIS

ABOUT THE AUTHORS

Cyril Bouquet and Michael Wade are Professors of Innovation and Strategy at IMD. Michael also holds the Cisco Chair in Digital Business Transformation and has published nine books. Jean-Louis Barsoux is a Research Professor at IMD and the co-author of several books on management, including the award-winning *The Set-Up-To-Fail Syndrome*.

A.L.I.E.N.
THINKING

*How to Bring Your
Breakthrough Ideas to Life*

Cyril Bouquet,
Jean-Louis Barsoux
and Michael Wade

BUSINESS

PENGUIN BUSINESS

UK | USA | Canada | Ireland | Australia
India | New Zealand | South Africa

Penguin Business is part of the Penguin Random House group of companies
whose addresses can be found at global.penguinrandomhouse.com.

First published in the United States of America by PublicAffairs 2021
First published in Great Britain by Penguin Business 2021
001

Print book interior design by Linda Mark

The publisher is not responsible for websites (or their content) that are not
owned by the publisher

Printed and bound in Great Britain by Clays Ltd, Elcograf S.p.A.

The authorized representative in the EEA is Penguin Random House Ireland,
Morrison Chambers, 32 Nassau Street, Dublin D02 YH68

A CIP catalogue record for this book is available from the British Library

HARDBACK ISBN: 978–0–241–48197–4
TRADE PAPERBACK: 978–0–241–48198–1

Follow us on LinkedIn: https://www.linkedin.com/company/penguin-connect/

www.greenpenguin.co.uk

Penguin Random House is committed to a
sustainable future for our business, our readers
and our planet. This book is made from Forest
Stewardship Council® certified paper.

To my dearest children, Rémy and Micah.
You remind me every day what matters most, and I will
forever treasure your love, happiness, and incredible ability
to step up to new challenges. You make me proud every day.

—CYRIL

To Astrid, Chloé, and Katarina for their energy and inspiration.

—JEAN-LOUIS

To my kids, both ALIEN thinkers, who, despite my best efforts to derail,
have turned into kind and thoughtful adults; and to my long-suffering
wife, Heidi, who surprises and delights me each and every day.

—MIKE

Contents

Introduction

"**Your job is to kill your own business**," said Amazon CEO Jeff Bezos. "I want you to proceed as if your goal is to put everyone selling physical books out of a job."[1]

Bezos was talking to Steve Kessel, leader of a small team of Amazon engineers who'd spent the last few years developing a new e-book reader. Toiling in an old law library in Seattle, surrounded by the very books they hoped to render obsolete, the team had made significant progress.[2] But something wasn't quite right. Their founding vision was beginning to feel small. There was nothing magical about the new device, nothing that really distinguished it from the competition. Then Bezos had an idea . . .

It's been more than a decade since Amazon unveiled the first Kindle. In fact, if you're reading the digital edition of this book, there's a good chance you purchased it from Amazon and wirelessly downloaded it to either your Kindle or another mobile device that supports MOBI, Amazon's e-book format. Today, Amazon so thoroughly dominates the market for electronic and print books that you may not remember a time when this wasn't so. But Amazon's preeminence was not a historical inevitability. In 2006, Barnes and Noble was the Goliath of book retailers and the e-book was an emerging technology struggling to gain acceptance with publishers and the reading public. And it appeared that Sony, not Amazon, was about to disrupt this industry.

In September of that year, Sony introduced the PRS-500 Portable Reader, a beautifully crafted e-reader that was hailed as "the iPod of the book industry," "the electronic gadget that could change the way we read."[3] Priced at $350, the Reader was slim and lightweight, and it achieved a feat that no previous e-reader had managed. The device featured an "electronic ink" technology that made e-book viewing as easy on the eyes as a printed book. It also had a brighter screen, longer battery life, and more memory than its competitors. To buy an e-book, you simply selected a title from one of the ten thousand available at Connect.com, Sony's online bookstore. Then you downloaded it (in Sony's proprietary BBeB format) by connecting the Reader to your PC with a USB cable.

Compared to every other e-reader in the fledgling market, the Sony Reader was avant-garde. In terms of design, it was a sleek and stylish Porsche among a bunch of jerry-rigged Model Ts.

It was also a massive flop. Within a few years, the Sony Reader was a historical curiosity.

To understand why this happened—and happened so fast—we must look at how Amazon's development team approached the task of creating the Kindle.

What consumers most wanted was an e-reading experience that approximated print. They also wanted all the advantages of digital media—instant access, ease of search, and storage of multiple books on one device—something like a portable bookstore. But portability was a sore spot with publishers. Having seen how peer-to-peer (P2P) file-sharing services like Napster had disrupted the music business, book publishers were anxious about embracing a technology that might allow customers to pirate their intellectual property with impunity. Then there was the question of profitability. The publishing industry wasn't quite sure how to make money by selling digital files instead of physical books.

Every e-reader development team faced the same conundrum. On the one hand, they needed a device that would make e-books profitable for publishers and prevent consumers from infringing on intellectual property rights. Unless this was done, publishers would have no incentive to modify their internal processes to make a critical mass of e-books available. On the other hand, consumers wanted not just a pleasant reading experience but a *purchasing* experience that was convenient, fast, and flexible.

Sony's fix for the portability problem was to force users to download every title from their PC with a USB cable. However, this was just one of many unproven solutions. And it wasn't enough to pry publishers away from their venerable business model. Besides, given the inconvenience of the Sony download process, many people simply preferred to keep buying hardcovers and paperbacks from their local bookstores.

At first, the Amazon team planned to build a device that also required consumers to download titles from a PC with a USB cable. But Bezos pushed back: "Here's my scenario. I'm going to the airport. I need a book to read. I want to enter it into the device and download it right there from my car."[4]

To achieve instant downloads, the device would have to be wireless. So Amazon partnered with Qualcomm to build a system called Whispernet, which would give Kindle users free 3G connectivity to download books from anywhere—almost instantly.

Compared to the Sony Reader, the first Kindle was unimpressive. It was larger, heavier, and had an inferior screen. However, it did solve the portability problem for publishers because users could download books only from Amazon. What was clearly a constraint for users was an advantage for publishers because it offered strong antipiracy protection. Amazon's sophisticated digital rights management system prevented users from sharing their books with friends or other devices, or even from connecting to a printer. At the same time, the purchasing experience itself was much faster and more convenient for users.

Just as important, Amazon solved the profitability problem by (initially) subsidizing e-book sales with profits from the Kindle device. Amazon e-books were a loss leader, but Amazon's hefty payments to publishers gave them the incentive they needed to crank out hundreds of thousands of Kindle titles.

Sales of the Kindle were explosive. Despite its aesthetic shortcomings and $399 price tag, the entire inventory sold out within six hours of its debut. From there, Amazon took the lead so quickly that every other competitor was left choking in its dust.

Amazon triumphed. And it triumphed with what was, in some respects, a weaker device. But as Bezos observed, "This isn't a device. It's a service."[5]

While Sony had engineered an incrementally better product, the performance gains were quite conventional because their new offering was itself the product of conventional thinking. Meanwhile, by reconceiving the entire publishing ecosystem, Kindle became the one-stop shop it is today—a mobile store that offers millions of people instant access to millions of affordable e-books. And the Sony Reader remains nothing but a footnote in the history of innovation.

THE BEGINNER'S MIND

ALIEN Thinking is not a story about Bezos, Amazon, or Sony. It's about the underlying reasons for innovation success (and sometimes failure). It's a study of what drives successful creativity and innovation, and it presents a model that can enable you to systematically generate one breakthrough solution after another.

Sony designed an elegant device, but Amazon devised a novel solution. It was less a battle of competing technologies than a contest between conventional thinking and original thinking. The orthodox approach works fine much of the time, but when you hit an impasse, it's often a sign that the solution requires the kind of divergent thinking demonstrated by Bezos.

To find untapped innovation opportunities and come up with radically new ways to tackle problems or aspirations, you need to see the world with fresh eyes and develop a state of mind known as "vuja de."[6] Breakthrough innovators avoid "déjà vu syndrome"—the feeling that we've experienced a situation before that tends to encourage incremental changes to old ideas. We argue in this book that all of us can develop new ideas and insights if we have strategies to challenge default assumptions, and if we notice the little curiosities that other people miss. As French novelist Marcel Proust famously observed:

> The only true voyage of discovery, the only fountain of Eternal Youth, would be not to visit strange lands but to possess other eyes, to behold the universe through the eyes of another, of a hundred others, to behold the hundred universes that each of them beholds, that each of them is.[7]

We must think and act as newcomers or outsiders if we want to notice and exploit innovation opportunities that are often sitting right in front of us. But while a growing body of literature offers rich insights into the nature of creativity, to date, there has not been an overarching framework to help individuals and organizations systematically provoke this shift in perspective—to help them step back from everyday routines and habitual behaviors, and steer innovations from early concept development to validated breakthrough solutions.

BREAKTHROUGHS ON DEMAND

What prompted Jeff Bezos to abandon the conventional approach to downloading e-books in favor of the wireless approach? What sparked that leap of imagination, and how can the rest of us mere mortals think more like Bezos?

If we could intentionally duplicate such a leap on a regular basis, we could exponentially increase the ability of individuals and organizations to develop amazing new products, services, and technologies. Such a mechanism might be like bottling a da Vinci or packaging a Steve Jobs—it could put an "innovation genie" at your command.

We have spent the past decade researching pioneering thinkers and change makers in a wide range of professions: entrepreneurs, designers, medics, architects, scientists, chefs, and artists, among others.

And our work with corporate clients at the Institute for Management Development (IMD) has ranged from running innovation workshops with social entrepreneurs and NGOs, to supporting family businesses in their efforts to change, to orchestrating massive transformation journeys with hundreds of senior executives in the commercial and government sectors. Through this work, we have identified a recurrent pattern in the way creative solutions evolve and survive. These patterns are captured in our ALIEN thinking framework. This framework is not merely an alternative mindset but also an operational model—an innovative thinking process that promotes the development and delivery of breakthrough solutions *on demand*. It can be used to come up with novel solutions to any given problem you may be facing in your life or in important aspects of your business.

Although it's important to understand the habits, strategies, tactics, and processes that support and nurture innovation, the ALIEN framework also organizes these principles into a mechanism that can be deployed again and again. Think of it as a device that helps cultivate original thinking.

ALIEN stands for Attention, Levitation, Imagination, Experimentation, and Navigation. As you'll see in the next chapter, unconventional thinkers focus their **attention** closely and with fresh eyes. At various times, they also step back and/or step away from the creative process to gain perspective and enrich their understanding, a process we have dubbed **levitation**. In addition, unconventional thinkers hone their ability to recognize hard-to-see patterns and to connect seemingly disparate dots. This allows them to **imagine** unorthodox combinations and to **experiment** quickly and smartly. Finally, they learn to **navigate** potentially hostile environments outside and within their organizations. This enables them to safely incubate their ideas and recruit powerful allies, so their ideas aren't shot down during the embryonic stage. The challenge in every part of this framework is to overcome biases and mental models that can constrain creativity or doom a great idea.

We don't offer a literal genie or bottled da Vinci. But with the ALIEN framework, we *do* offer a set of innovation tools that can be applied to any endeavor. No longer do you need to wait for inspiration to strike—if it ever does. You can use the ALIEN model to catalyze original thinking and fast-track your ability to spot patterns and make the right mental connections.

Although our focus is on individual managers and professionals, the ALIEN framework applies equally well to teams and organizations of all kinds and sizes who want to smash the logic of orthodoxy and find novel responses to increasingly complex problems. We find ALIEN thinkers in business, in architecture and design, in the arts and sports, in science and social entrepreneurship. So while the approach we present in this book will often appeal to innovation professionals, it contains valuable insights from, and for, leaders in many walks of life.

What they, and you, have in common is a desire to push through the veil of orthodox thinking to more consistently uncover new solutions to some of the world's most vexing and intractable problems. We're ready to show you exactly how the ALIEN thinking model can help you to accomplish just that.

ONE

Discovering the DNA of Originality

I N MAY 2014, BILLY FISCHER GOT A CALL FROM THE WORLD Health Organization (WHO). Fischer, a pulmonologist and critical care physician, was asked to move to Geneva for two months to provide technical expertise on severe viral infections. Thinking he'd spend most of his time poring over technical documents, he agreed. Seventy-two hours later, he flew to Geneva, but when he walked into the WHO office, he learned there'd been a change of plans. The deadliest Ebola epidemic in recorded history had just begun, and WHO officials wanted him to go immediately to Guinea, in West Africa, to help the Doctors Without Borders fighting the disease.

Fischer is the first to admit that he was scared to go. Ebola is a highly contagious disease that causes severe internal bleeding and is very often deadly. The village to which he was traveling had a clinical mortality rate of more than 90 percent. "We were losing," he said. "They hadn't had a survivor in months."[1]

In addition, Fischer wasn't sure he was well suited to join the Doctors Without Borders team. Though he did have experience working in resource-constrained environments, he had never treated this particular virus before. Nevertheless, he agreed to go. He had to try to help.

As it turned out, Fischer's outsider status proved to be a lifesaving asset.

"EXTREMES OF CREATIVITY"

When he arrived at the village in Guinea, Fischer saw that the term *resource constrained* had acquired a whole new meaning. He had almost none of the tools he normally relied on to diagnose and treat patients, and he couldn't use others because he had to wear a protective suit that covered him from head to toe. "I didn't have CT scans. I didn't have X-rays. I didn't even have labs. I didn't use a stethoscope because I couldn't put it in my ears. I'm reduced to my eyes and to tactile sensation." He had to pay attention to different symptoms than he normally would have and to think differently. "What I found is, if you do that, you can actually gain a lot."[2]

Some treatment facilities were nothing but tents filled with cots for the sick and dying. And some of the equipment they did have wasn't functioning. "We had these handheld [diagnostic] machines that could give you basic clinical chemistries. You could draw somebody's blood, put it into the machine, and two minutes later, you had a full report. Problem was: the machines weren't designed for tropical climates. They were designed for 15°C [59°F] and air-conditioning, with no humidity. The humidity caused a breakdown of the machines. Also, you had to regularly update the software. We're in a place with limited electricity and almost no internet connectivity, but the machines would stop working if they weren't updated regularly."

Instead of updating the software, Fischer and his team changed the dates on the machines to an earlier day and time. This tricked the devices into thinking that it wasn't time for an update. "We went to extremes of creativity to fool these high-tech things."

"Another thing we did was figure out how to use bowls of rice to act as a desiccant [dehumidifier] to suck up moisture so the machines wouldn't break down. We used little dorm refrigerators to keep them cool. We'd keep the refrigerators open and operate the devices inside the refrigerator when we needed them."[3]

By turning off his learned responses and instead encouraging his team members to propose off-the-wall solutions to tricky problems like the devices' malfunctioning, Fischer unleashed their collective ingenuity. As a result, a number of standard work routines were completely revamped. One such routine involved communications between high-risk zones, where cumbersome protective gear was necessary, and low-risk zones, where the suits weren't required. Because doctors couldn't quickly move between the zones, and because it was almost impossible to talk with someone while wearing a "space suit," it was difficult to communicate confidential medical information between the zones. Everything that went into the high-risk zone was supposed to be destroyed. To facilitate better communication, the team came up with the idea of using their iPhones (which could be easily sanitized) to take photos inside the high-risk zone and then wirelessly transmit them to a printer on the low-risk side, about six meters away. It worked!

Ideas like this were often brainstormed during mealtimes, when team members proposed and debated one idea after another until a solution to the latest challenge was devised.

A SHIFT IN ATTENTION

Still, despite the team's best efforts, patient after patient died in the makeshift quarantine and treatment facilities. Desperately ill patients no longer viewed doctors and nurses as potential saviors but as villains—as harbingers of death. The prognosis for anyone contracting the disease sweeping the village was so dire (90 percent of patients didn't survive) that families would hide their loved ones from caregivers—and risk contracting the deadly virus themselves—rather than let a medical team take them to the treatment center. Better to die among family and receive a proper funeral than to perish among strangers garbed in ghoulish hazmat suits. This fear threatened to spread the disease even farther and faster.

For decades, the WHO's approach to an infectious disease had been to focus first on stopping its spread: identify the infected person, isolate that infected person, and track people who had any exposure to that infected person. Taking care of the patients who were already infected was less important. This historical approach to Ebola was no longer working—and Fischer pointed this out.

To their credit, said Fischer, the organization decided to enlist more outsiders—people not constrained by the established paradigm—to combat the epidemic. Along with specialists in infectious diseases, they also recruited people who specialized in critical care (like Fischer). These doctors were given two tasks: (1) learn about the clinical course of the illness to develop better patient care, and (2) produce better patient outcomes by testing new treatment combinations.

Fischer began by persuading his colleagues to shift their focus from containment to observing the progression of the disease. They found that, during the first two to three days of infection, people exhibited nonspecific symptoms, which made it extraordinarily difficult to identify an infected patient. Patients had a fever, headache, weakness, fatigue, and a sore throat, all of which could be attributed to any number of other ailments. After about three to five days, though, the symptoms became unmistakable. People developed nausea, profuse vomiting, and severe diarrhea—sometimes losing more than ten liters of fluid (more than the entire blood volume of a human being).

Fischer noticed that, around the beginning of the second week of infection, either the severe diarrhea led to dehydration that produced multiorgan failure (and then death in most cases), or the patient's immune system began to control the virus's replication. If the latter occurred, patients lived.

Although the literature on Ebola contained references to diarrhea, "nobody ever talked about how *much* diarrhea they had," said Fischer.

This was probably the most impressive thing that caught our attention, but it was also an opportunity. This was something that we could reverse on the ground. We could replace the volume loss with fluids. And so this became a challenge; this became an interesting experiment for us.

What we ended up doing was, everyone who was confirmed with Ebola received an IV and got aggressive fluid resuscitation. We had to guess on how much electrolytes they were losing—the potassium and magnesium that are in sports drinks, you lose those in your diarrhea—and so we were guessing. . . . I would basically look at a container of diarrhea, slosh it around, try to figure out how much they'd lost, and then respond to how much potassium I thought they'd lost. And then we gave everyone antibiotics and we gave everyone antimalarial medications.[4]

With these three simple interventions, the doctors were able to reduce the mortality rate among Ebola sufferers in the village from over 90 percent to less than 50 percent.

"We were astounded," said Fischer. "This was a critical moment because it showed us that Ebola is not a uniformly fatal disease, but that the supportive care that is a cornerstone of therapy for every other severe viral infection also works on Ebola. It sounds obvious, but it was fundamentally groundbreaking to show that with aggressive supportive care, we could reduce mortality."[5]

Within weeks of the changes, the team accomplished its goals. They reduced community resistance to quarantine by inspiring hope instead of fear in the villagers. In turn, this helped halt the transmission of Ebola to other areas of Guinea, as well as the neighboring countries of Liberia and Sierra Leone.

Because Fischer didn't specialize in the treatment of Ebola, it was easier for him to forgo the traditional quarantine approach and adopt a radically different strategy. His outsider status helped him assess the problem with fresh eyes and with fewer preconceptions of how he ought to deal with the specifics of the crisis.

A VERY DIFFERENT CRISIS

Two years earlier and half a world away, Teresa Hodge had faced a very different crisis. She was sitting at her computer, filling out an application for a work-at-home job, when she encountered a question that sent a jolt of fear through her body. She paused, took a deep breath, and typed, "Yes."

The screen went blank.

Then a message she'd been dreading appeared on the screen: "We're sorry, but your response to one of the questions indicates that you are not qualified for this job. Thank you for your time."

Teresa was saddened, but not surprised or confused. She knew why the process had so abruptly ended. She was one of the millions of US job applicants who are automatically rejected for employment each year when they answer "yes" to the question "Have you ever been convicted of a crime?"[6]

Although the US accounts for just 5 percent of the world's population, it houses 25 percent of its prison population. As of 2020, the American criminal justice system holds almost 2.3 million people in 1,833 state prisons, 110 federal prisons, 1,772 juvenile correctional facilities, 3,134 local jails, and 218 immigration detention facilities, as well as military prisons and other facilities.[7] Once their sentences have been served, many of these former prisoners find it extremely difficult, if not impossible, to find good jobs.

Eight out of ten employers now run criminal background checks on job candidates, and if someone has an arrest or conviction record, it typically triggers an automatic rejection. Small wonder, then, that of the six hundred thousand plus people who leave prison every year, more than half are still unemployed twelve months later, and 70 percent eventually end up back in prison, according to Hodge. To make matters worse, the high rate of incarceration and unemployment affects not only the former prisoners but also their spouses, children, other family members, and entire communities, locking them into a never-ending cycle of imprisonment, recidivism, and poverty. Hodge dubs this the "invisible life sentence."[8]

Ironically, Hodge was in the midst of attacking this systemic problem when her employment application was rejected. She was applying for a job to pay some bills while she laid the foundations for a nonprofit organization called Mission: Launch, which aimed to help people with arrest and conviction records to access loans and start businesses.

After a disgruntled ex-employee made accusations to the government about a company that Hodge had cofounded, she was eventually convicted of mail fraud (a crime she denies) and served a seventy-month sentence at Alderson Federal Prison Camp in West Virginia (a facility that once housed "doyenne of domesticity" Martha Stewart and jazz singer Billie Holiday).

"I never thought that I would find myself incarcerated under any set of circumstances," said Hodge. "It was just very devastating to me and my family. I was fearful of the women I would encounter. What was shocking, however, was how much I had in common with the women I was incarcerated with. I went from fear to feeling a part of a community of women who were far away from their homes, and many of them, like me, just wanted to get back to their lives."[9]

During her incarceration, Hodge was far from idle. She continued to develop her business skills by reading books on entrepreneurship and reviewing business plans for friends and fellow inmates. She also read articles about the plight of inmates like herself. For example, she learned that 25 percent of children in the US had an incarcerated parent, and that three in ten of those children would wind up in prison themselves.

She also had plenty of time to worry.

I sat in prison with a sinking feeling, knowing that technology was transforming the world, while I was left behind. I arrived in prison on January 3, 2007. Six days later, Steve Jobs, the founder of Apple, stood on a stage and presented the device that would change the world. He called it a 3-in-1 product "a widescreen iPod with touch controls, a revolutionary mobile phone, and a breakthrough internet communications device." He was correct; the device, the first generation iPhone, changed everything. I had the frame-of-mind to know what that meant in my life and what it meant to the future of my ability to work and keep pace with society. Tech advances and a changing society in so many ways means that individuals living with criminal records are facing more-and-more significant financial inclusion barriers.[10]

In addition to reading, Hodge filled her days by talking with fellow inmates to learn their stories. What were their dreams? What crimes had they committed, and why? What kinds of family backgrounds did they have? One of her key observations was that many of the women who were released quickly ended up back in prison. She wanted to know why. The answers were usually the same: the women had no family, or had unsupportive family, and were unable to find a job after release.

As time passed, Hodge internalized these inmate narratives about crime cycles, inescapable poverty, and the hunger for new opportunities. She came to understand why so many people were coming back to prison. She came to understand that "if you can't get a job, you can't get on your feet."[11] (She also realized that she'd been blessed with a very supportive network of friends and family, which made her feel "privileged" for the first time in her life.)

Going to prison was the easy part. Coming home was much harder.

"When lying on my bunk in prison, I dreamt of being able to launch a business; my goal was to empower not only me but so many formerly incarcerated women."[12] So she started thinking: "What will I and others need?" "What are the right services" to help former prisoners "reconnect back to society in a meaningful way?"[13]

She didn't wait for release from prison to pursue her dream. She began brainstorming ideas with her daughter, Laurin Leonard, in the prison visitors' room. By the time Hodge was released in 2012, she was ready to kick her new venture into high gear.

Hodge realized that making a tangible impact on the lives of former prisoners would require pioneering efforts outside of the conventional public policy framework. So she and her daughter created Mission: Launch. As a former entrepreneur, Hodge proposed solutions that focused on business development and entrepreneurship, helping people who couldn't find suitable jobs turn their skills (e.g., catering) into a business.

This entrepreneurial focus was less a choice than a necessity for people who couldn't get hired. It was also a creative way of bypassing the problem of reemployment by helping those with arrests or convictions become self-employed. Beyond this reentry strategy, Mission: Launch also teamed with government, the private sector, and community development financial institutions (CDFIs) to champion social-good technology and civic innovation.

In the process, the mother-daughter team did the unprecedented: they connected a diverse group of criminal justice stakeholders (from civil servants and private companies to lawyers, former prisoners, and activists) to rethink criminal justice policies. Mission: Launch sponsored and developed several initiatives—from Fair Chance Employment (enforcing fair-hiring legislation) to Clean Slate DC (sealing criminal records). The organization also initiated a series of two-day collaborative events known as "hackathons" where programmers and data scientists, entrepreneurs, designers, engineers, and policy makers could create fast solutions to the challenge of rebuilding reentry. In 2015, Mission: Launch began a sixteen-week "entrepreneurship boot camp," and its start-up accelerator won a $50,000 prize from the US Small Business Association, which helped attract matching funds from other foundations and companies.

Hodge became good at cultivating allies and raising the profile of her organization by serving as a frequent panelist and speaker (her speeches include a 2016 TEDx Talk), leveraging her firsthand experience of prison to win support. On the strength of that exposure, she forged a partnership with singer-songwriter John Legend and his FreeAmerica initiative, and another with Bank of America.

A DATA-DRIVEN SOLUTION

Another major turning point for Hodge in 2016 came when she received a request from a community-oriented bank to vet a loan candidate with a criminal record. "They didn't know how to assess her," she said.[14] For the next three days, Hodge and her daughter searched for existing tools that could help with the assignment. They ended up finding nothing. But instead of giving up, they decided to create a list of factors that a lender should consider before granting or denying a loan to someone with a criminal record. Hodge had a pretty good idea of what was needed, and within a few days, her team had created a framework.

Her team's analysis revealed that the applicant in question was a credit risk, but it also taught Hodge a valuable lesson. "No one is going to spend three days trying to vet a single person," she said. "I recognized that it would be just easier to say no. But no one should be a no forever."[15] With that insight, she decided there might be a market for a new technology platform that could perform such an analysis for customers, much the way employers, banks, and others use outside vendors to run credit checks on applicants. Soon their work produced a ripple effect as other community lenders approached her with similar requests, so Hodge persuaded a software engineer to build a financial technology tool that would perform the kind of analysis her team had just done.

The result was R3 Score, a risk assessment tool powered by an algorithm that assessed a person's criminal history, along with volunteer work, education, credit history, employment experience, and information self-reported by individuals to de-risk decisions for prospective employers, banks, and landlords. R3's software produced a numeric score that predicted future trends. The scores, which deliberately mimicked the FICO scores credit bureaus use to rate people's credit histories, ranged from 300 to 850. The higher the number, the less risky the person.

By providing employers, landlords, and financial institutions with a more nuanced, data-driven risk evaluation from a third party, Hodge hoped that she could enable them to make more intelligent decisions about whether someone posed a good credit, housing, and employment risk.[16] R3 Score was an alternative to the reports delivered by traditional background companies, which are composed entirely of criminal records. "While it's important that we discontinue the practice of disqualifying applicants for simply having criminal records, it's also imperative that HR professionals have access to other criteria they need to determine a candidate's eligibility," said Hodge. "Formerly incarcerated people need a way back into the workforce, and HR professionals need a way of weeding them in instead of out of the pile."[17]

Beta testing with banks in 2017 confirmed the value of R3 Score, and Hodge and her team also worked with two CDFIs and a start-up to better understand their needs. As of this writing, R3 Score has signed on five contract partners to help fund entrepreneurs with criminal records.

More recently, after the COVID-19 pandemic caused unemployment to skyrocket and reduced the goodwill of businesses, Hodge decided to pivot. Without completely writing off the B2B market, R3 Score has shifted to a B2C model, targeting consumers directly rather than corporations, CDFIs, and other business clients. The goal was to offer a "freemium" model, funded with money raised from philanthropic sources and investors, aimed at delivering ten thousand scores into the hands of customers by the end of 2020.

It's still early, but R3 Score has the potential to become a real game changer by improving social outcomes for the marginalized—for the millions of people living with criminal histories.

"We shouldn't be allowing employers to discriminate against millions of Americans, keeping them locked into sentences and mistakes . . . they did five, ten, fifteen, twenty, thirty years ago," said Hodge. "We need people who are coming home from prison to have their human capital restored so that they can be good mothers and fathers, good community members, and take care of their families."[18]

WHAT'S SO ALIEN ABOUT THIS THINKING?

What separates ALIEN thinkers like Billy Fischer and Teresa Hodge from the rest of us? How did Fischer and his team connect the dots from Ebola's clinical progression to aggressive fluid resuscitation to improved outcomes and a greater willingness among villagers to submit to quarantine? In retrospect, the strategy seems obvious. But if it was so obvious, why did no one else have these insights?

Why did no one, before Teresa Hodge, think to create a more-nuanced risk assessment tool for people with criminal records? It's no secret, at least among employers, real estate professionals, and financial institutions, that most people with arrest or conviction records are automatically (and often unfairly) denied the opportunities they need to reintegrate into society.

The answer is that ALIEN thinkers (intuitively or consciously) apply the ALIEN framework, which comprises **attention**, focusing hard on a particular context or population to understand its dynamics and latent needs; **levitation**, stepping back to reflect on the situation and make sense of it; **imagination**, envisioning that which is not and generating avant-garde ideas; **experimentation**, turning a promising idea into a workable solution that addresses a real need; and **navigation**, adjusting to the forces that can make or break your solution.

By looking at the cases of Fischer and Hodge, we can tease out three key personality traits of ALIEN thinkers.

Rebels with a Cause

ALIEN thinkers question what others take for granted. For example, Fischer noticed and dared to point out that the existing approach was contributing to the spread of Ebola rather than helping to contain it. Likewise, Hodge challenged the model of traditional background check companies whose reports consisted entirely of arrest and conviction records. But Fischer and Hodge didn't defy conventions just for the sake of it. They disrupted the established order to tackle causes that matter. ALIEN thinkers are unconventional, but they're also respectful of the broader purpose that motivates their search for progress. They are committed to the idea of creating meaningful change in business and society. They fully understand the danger of being misperceived as renegades or loose cannons. Both Fischer and Hodge maintained a very clear focus on what they were trying to do and on those they sought to serve.

Curious Integrators

ALIEN thinkers are people-centered, trying to understand what drives others and to make sense of their worlds. Both Fischer and Hodge learned by taking an interest in individual cases and interacting with people who were struggling. But both were also system-minded and able to take into account the larger issues and multiple stakeholders involved. For example, Fischer first tried to make sense of the entire process of patient care before trying to tackle the cycle of Ebola contagion in different ways. And Hodge rethought the whole approach of risk evalutaion vis-à-vis former prison inmates. She then created an easy-to-understand score for ex-inmates—one modeled on the FICO score with which institutions were already familiar and that many were already using to assess candidates for loans and employment.

Ingenious Analysts

ALIEN thinkers combine their creative intelligence with whatever analytic tools they have at their disposal. Even in a resource-constrained environment, Fischer and his team demonstrated considerable creativity, using rice and mini refrigerators to maximize the effectiveness of what little technology they had and extend its performance boundaries in a hostile environment. On a larger scale, Hodge embraced technology to quantify people's personal and institiutional histories, and to do predictive work. She used technology to augment her innovation capabilities and develop original solutions to the seemingly intractable problems afflicting the American justice system.

WHY NO ONE WANTS TO BE THE ALIEN THINKER

The cases of Fischer and Hodge show how an ALIEN thinker mindset can lead to breakthrough solutions.

In the early 2000s, futurists Edie Weiner and Arnold Brown argued that ridding ourselves of preconceived notions about the existing ways of doing things can free our minds and stretch the boundaries of our perception.[19] Overfamiliarity makes it difficult to approach the future in innovative ways. Aliens from another planet would be better at seeing the world from an unbiased perspective. They are not prisoners of the assumptions, prejudices, and recipes that constrain our insights and ingenuity.

The idea of an alien looking at our world without bias was an eye-opener for us, providing the initial spark for this book. To improve our ability to imagine how things can change, we must adjust our lenses to see as an alien would see.

But developing this shift in perspective isn't easy.

To begin with, there are many psychological barriers to ALIEN thinking: biases that limit our ability to notice things, our creativity, and our willingness to switch course. It's not always easy to be the alien in the room. Most of us have a natural conformity bias, making it difficult to be the misfit who always has a different outlook on things. Another particularly damaging bias is our tendency to experience more regret when a bad outcome is the result of trying something new than when a bad outcome is the result of sticking to our usual approach.[20] In other words, we punish ourselves for introducing innovative ideas that don't work out—for thinking like aliens.

Finally, even if we *do* manage to come up with original ideas or potential solutions, we never know how they will be received. Science fiction novels and films have taught us to be wary of aliens—not because they are always bad, but because we never know their intentions. The same distrust extends to ALIEN *thinkers*. If you can't assess the probable impact of a new idea, you are likely to treat it and its initiator with suspicion. A fascinating example is the case of James Dyson and his bagless vacuum cleaner. Several manufacturers—including Philips, Electrolux, Black and Decker, and Hoover—turned down the prototype because it threatened their lucrative sales of replacement bags. More telling still was the subsequent remorse of Hoover's then vice president for Europe: "I do regret not taking the product technology off Dyson. It would have lain on the shelf and not been used."[21]

This reaction captures the forceful pull of the status quo. It's perfectly understandable that organizations seek to protect themselves from novel ideas. No one knows in advance whether an original idea will boost performance or weaken it. But in this case, the implication is that the organization would rather prevent a successful product from coming to market than adopt it if it would disrupt its business model and industry paradigms. The corporate immune system includes a number of forces that mobilize themselves to reject inconvenient ideas and block uncomfortable notions of progress.

Resistance to ALIEN thinking and ALIEN thinkers is ubiquitous. So your dual challenge is learning how to think like an alien without falling prey to the systems of resistance that might hamper your efforts to change.

Our framework helps you do both.

THE ELUSIVENESS OF BREAKTHROUGH SOLUTIONS

The digital advances of the past two decades have enabled a much broader population than ever before to express creative intelligence.[22] Unconventional thinkers the world over have unprecedented access to the distributed knowledge, talent, capital, and consumers they need to create a start-up or a movement around a great idea. Innovation has been thoroughly democratized.

And yet breakthrough solutions remain hard to come by. Apart from the internet-driven service disruption, we have not seen spectacular surges of innovation across sectors. Respected economists, including Tyler Cowen and Robert Gordon, have spoken of "innovation stagnation."[23] Leading business thinker Gary Hamel noted that corporations are awash in ideas that fall into two buckets: incremental no-brainers and flaky no-hopers.[24] In our work with innovation teams, we see many promising ideas that either perish or end up as superficial, narrow, or skewed solutions.

This lack of progress is surprising given our improved understanding of the innovation process, driven in large part by design thinking and lean start-up methodologies. Terms such as *user-centered, ideation,* and *pivot* have become commonplace and have changed the way innovators and organizations think about creating new offerings. Yet for all this guidance, only 43 percent of corporations have what experts consider a well-defined process for innovation, according to the research firm CB Insights.[25] And companies are slowly but surely losing confidence in their ability to innovate.

When we talk with entrepreneurs and executives about existing innovation frameworks, their criticisms center on three overlapping issues. The ALIEN framework addresses each issue.

They're Unrealistic

The still-influential waterfall or stage-gate approaches, for example, are overly linear, showing little regard for the constant zigzagging between activities that may be needed.[26] Elmar Mock, the serial entrepreneur who invented the Swatch watch, put it this way in a recent podcast: "The very natural instinct for an innovator is to move in a nonlinear way, to go from concept to know-how back to concept, to re-question your concept, to re-look for new know-how, to change your concept again."[27]

Design thinking—with its emphasis on understanding user needs, encouraging divergent thinking, and promoting the iteration of ideas through constant trial and error—offers more flexibility. It recognizes that innovation is nonlinear and that it often evolves through multiple feedback loops. Unfortunately, in design thinking these loops typically follow well-worn paths, from testing back to prototyping, brainstorming, or even redefining the problem, whereas innovation happens in much more haphazard, accidental, and fortuitous ways.

ALIEN thinking, by contrast, offers freedom within a frame. It captures the fluid way that many historical breakthroughs have emerged. It accepts that innovation can start anywhere, follow any pathway, run concurrently, and loop back repeatedly. The only rule is that you must devote time to each of the five ALIEN components for the innovation to succeed.

They're Incomplete

Existing models don't fully recognize the digital aspect of innovation or show how technology can augment the influential "human-centric" principles enshrined in design thinking. Digital technology offers many additional ways of gaining empathy by tapping into the end-user's experience without necessarily going into the field and standing in other people's shoes. For example, COVID-19 created situations in which close observation of user behavior was either unsafe or unethical. Fortunately, innovators were able to use apps, sensors, and other digital technologies to study user habits and notice interesting patterns of behavior that might have been opaque to the human eye.

The executives with whom we've talked also questioned the prevailing emphasis on action and fast iteration (pillars of the lean start-up concept) because it ignores the creative impact of reflection, which is often misconstrued as procrastination. Finally, existing models seldom offer guidance on how one can manage emotions during an innovation journey. ALIEN thinking takes the innovator's psychology into account. At times, the search for breakthrough solutions creates anxiety, confusion, and discouragement. At other times, every signal is positive and your energy is high. Existing models don't do much to help innovators navigate this constant swirl of emotions. No wonder so many innovative efforts get derailed.

They're Misleading

Existing models gloss over the various pitfalls and biases that constrain original thinking, especially in organizational contexts. For example, by insisting on the need to immerse yourself in the world of users, design-thinking approaches neglect (or at least downplay) the role of other stakeholders in the innovation process, as well as the need for creativity in mobilizing their support for a novel solution.

Executives know that to devise ingenious solutions, they have to break paradigms and shift mindsets, but they usually lapse into standard ways of thinking when it comes to delivery. The failed Sony Reader is a case in point. All the creativity that went into developing that sleek solution was undone by a lack of originality in execution. Sony neglected to line up the book publishing industry as an ally—a mistake that Amazon did not make when it launched a technically inferior product that made it easy for users to download e-books.

For your stellar solutions to thrive, you need to approach unconventional partners (as the WHO did when it approached Fischer), identify underutilized channels, and invent new business models.[28] You need to invest as much creative energy in introducing and delivering the offering as you do in generating it.

Overall, existing innovation frameworks are too rigid and often difficult to recall in real time. They also ignore the real-world risks of challenging orthodoxy—particularly when you are part of an organization—namely, that you'll be ignored, ostracized, transferred, or fired.

The ALIEN framework fully recognizes the resourcefulness and ingenuity needed, at what is dismissively dubbed the "implementation" phase, to turn a workable solution into a success.

AN ANTIDOTE TO ORTHODOXY

ALIEN thinking provides a blueprint to escape conformity and to devise truly breakthrough solutions to important problems. There are five strategies that must be considered in turn: attention, levitation, imagination, experimentation, and navigation (ALIEN).

To illustrate the framework, we return to the cases of Billy Fischer and Teresa Hodge.

Attention is the act of focusing hard on a particular context or population to understand its dynamics and latent needs. By listening to villagers' fears of the Ebola quarantine—the long-established way of handling an outbreak—Billy Fischer came to understand that only better patient outcomes would induce them to enter quarantine (and treatment) instead of avoiding the physicians altogether. In a resource-constrained environment, and hampered by protective clothing, he also learned to pay closer attention to alternative symptoms using his senses of sight and touch. In Hodge's case, she paid attention to the recidivism among the inmates at her facility and asked them questions about why they had been imprisoned once again. The answers she received from the women triggered her search for a solution.

Levitation is the act of stepping back or "decentering" in order to expand and enrich your understanding. Having gathered insights into the realities of a situation, need, or challenge, you need space to make sense of those findings, work out what they mean, and prime your mind for inspiration. Fischer and his colleagues used their mealtimes to step back and gain perspective, which is why many of their best ideas were generated during breakfast, lunch, and dinner discussions. Hodge reflected on the stories her fellow prisoners told her about their lives, and also read articles about the underlying causes fueling the cycle of imprisonment and poverty among the inmates and their families. She pondered ways to break the vicious cycle until she came to understand that what ex-inmates needed most was access to opportunities—employment, housing, and financial opportunities that were often closed to them because of their criminal histories.

Imagination is the act of envisioning that which is not and then generating avant-garde ideas. Though often shrouded in mystery, imagination is essentially the result of making creative connections between existing concepts—joining the dots in new and interesting ways. Fischer imagined a quarantine unit that offered genuine hope rather than the prospect of death in isolation, and he drew inspiration from therapies that worked for other types of severe viral infections. For her part, Hodge imagined a nonprofit that would work with financial institutions and community organizations to help formerly incarcerated people access loans and start their own businesses.

Experimentation is the act of turning a promising idea into a workable solution that addresses a real need. Fischer experimented with untried combinations of aggressive rehydration, antibiotics, and antimalarial medication. And he tried giving plasma from Ebola survivors to patients fighting the disease in an attempt to boost their immune systems.[29] Hodge's efforts to vet a single loan candidate caused her to realize that there might be a market for a technology platform that would perform such analyses, prompting her to get a software engineer to build the financial technology analysis tool that became R3 Score.

Navigation is the act of adjusting to forces that can make or break your solution. Your belief in the solution (and overfamiliarity with the context) can lead you to underestimate the effort needed to mobilize supporters and steer past obstacles. Fischer had to persuade the authorities at the World Health Organization to change the protocols for dealing with Ebola. And he encouraged exchange of best practices with other NGOs in the region: "There was sharing back and forth. At one point we showed them what we were doing with our refrigerator. We would talk about different strategies that we were using to get around difficulties."[30] Hodge worked diligently to recruit individuals and organizations that would support the work of Mission: Launch and R3 Score, participating in conferences, demo days, hackathons, and other events to raise awareness of problems and her proposed solutions. Instead of trying to fight entrenched systems head-on, she often went around systems to help ex-offenders better help themselves through self-employment.

Unconventional thinkers need a simple, shared, and overarching framework for developing breakthrough solutions. They need guidance to escape the various psychological barriers that constrain their creativity, to leverage the digital tools that can boost their intelligence throughout the process, and to avoid getting shot down as they begin to think and act in truly original ways that threaten the standard (and safe) way of doing things. It is to address these challenges that we developed the ALIEN thinking model. The five components are presented in a sequence that makes intuitive sense, but many successful inventions unfold in unpredictable ways, and you can actually proceed in any order, leveraging each component to enrich any phase of your process.

The framework covers familiar themes—like paying attention to the world we live in, boosting your imagination, and experimenting intelligently. None of these are particularly revolutionary on their own. It's *how* they are implemented that changes everything. ALIEN thinkers think differently and are able to find effective, original solutions to almost any problem they face.

⊢═◇═⊣

KEY TAKEAWAYS

- ALIEN thinkers, such as Billy Fischer and Teresa Hodge, are rebels *with* a cause. In addition to questioning what others take for granted, ALIEN thinkers are:
  - » Curious integrators, trying to understand what drives others and to make sense of their worlds
  - » Ingenious analysts, combining their creative intelligence with the analytic tools at their disposal
- Although an ALIEN mindset can lead to breakthrough solutions, developing this state of mind isn't easy. There are many psychological barriers to ALIEN thinking—biases that limit your ability to notice things, your creativity, and your willingness to switch course.
- As illustrated by the case of James Dyson and his bagless vacuum cleaner, ALIEN thinkers and ideas are often regarded with mistrust by people with a stake in the status quo.
- Despite the digitally enabled democratization of innovation during the last two decades, breakthrough solutions remain elusive, in part because existing innovation frameworks tend to be unrealistic, incomplete, misleading, or a combination of the three.
- An antidote to orthodox thinking, the ALIEN framework consists of five strategies: attention, levitation, imagination, experimentation, and navigation (ALIEN).

QUESTIONS TO ASK YOURSELF

1. In your circles, who would you describe as an ALIEN thinker and why?
2. Can you recall a simple innovation framework without going online?
3. What's your typical gut reaction when you encounter original new ideas?

TWO

Attention
See the World with Fresh Eyes

A S A RURAL DEVELOPMENT RESEARCHER AND NATIVE OF INDIA, Narayana Peesapaty had noticed that for years there had been a groundwater shortage affecting more than half of the country's population, despite no change in rainfall. Over time and with further research, he began to unravel the mystery. It turned out that Indian farmers pay almost nothing for electricity, thanks to government subsidies, which incentivize them to constantly run their water pumps. As a result, India extracts more groundwater than the US and China combined.[1]

But this discovery led to yet another mystery. Why wasn't this water use, combined with cheap electricity, producing more crops and higher incomes for Indian farmers? More water and low overhead should generate more profit. But that wasn't happening. Instead, many farmers were struggling to stay afloat.

After digging deeper, Peesapaty finally pinpointed the culprit: rice. Although there had been no increase in India's rice consumption, more and more land was being added to rice cultivation every year and taken out of sorghum (a type of millet) production.

Rice is a thirsty crop. "It takes 5,000 litres of water to cultivate one kilo of rice," Peesapaty observed. "For one tonne of rice, you require five million litres of water. Ironically, every year, thousands of tonnes of rice go to waste in warehouses across the nation." Meanwhile, millet farmers were suffering because "other cash crops were killing their motivation to continue with their crops and also the ground-water levels [were getting] dangerously low."[2]

THE SOLUTION: EDIBLE SPOONS?

Peesapaty did something unusual. The typical response of a science consultant or researcher might have been to pressure the authorities to direct farmers toward less resource-intensive crops like millet. That's what researchers do. They try to persuade the powers that be to take action.

But Peesapaty had always been a rebel. He reasoned that if market forces had created the problem, market forces could solve it. And he wouldn't wait for someone else to tackle the challenge; he would do it himself. He would find a new use for millet—a grain that consumes sixty times less water than rice. The only question was, How?

Initially, Peesapaty took a conventional approach, trying to think of millet-based food products that would excite the public's taste buds. Breads and breakfast cereals came to mind (Millet Flakes, anyone?), but he quickly shelved those ideas. Nobody would buy those products.

It was time for some outside-the-box thinking.

"During a flight from Ahmedabad to Hyderabad, I was served refreshments in food-grade plastic. Holding the spoon, I was thinking about the amount of waste generated, considering the number of flights and passengers in each. My eureka moment was when I had a flashback of a field visit, in which we used the hard jowar roti [a flatbread made from millet flour] to scoop out the curry and dal. I toyed with the idea of using jowar to make a three-dimensional spoon."[3]

Before long, Peesapaty was doing more than just toying with the idea. He quit his job and launched Bakeys Foods Private Ltd. to manufacture a line of edible cutlery made from millet flour. If the concept succeeded, it would strengthen demand for millet. In turn, this would convince more farmers to cultivate the highly sustainable crop, reducing groundwater and electricity use, and cutting the amount of nonbiodegradable plastic destined for the nation's landfills.

Of course, this was all easier said than done.

For nine years, Peesapaty went without an income as he struggled to develop an edible spoon that was durable enough to withstand hot soup or coffee but soft enough to be eaten by consumers after finishing their meals. He also wanted to give the spoons a pleasant taste (in sweet *and* savory flavors) so that they could be paired with any meal and users would be less likely to throw them away.

It took a long time to perfect the baking process, and even longer to find a market for the spoons. At one point, Peesapaty even tried selling them outside grocery stores and parks, but few people were interested. To finance the business, he'd sold his houses in Baroda and Hyderabad and moved his family to an apartment. The money went to R&D, but by 2016—more than a decade after launching the company—he had little to show for his investment. And his financial situation had become precarious, with the bank even threatening to take his apartment.

Things were looking bleak for Peesapaty when a camera crew arrived from Better India, a website that focuses on positive news, to shoot a video about Bakeys. When the clip went live, his edible spoons became an overnight sensation. Emails began flooding in at the rate of one per second. Every time Peesapaty finished answering a phone call, he'd discover that dozens more callers were trying to get through. The video got five million views during the first week alone, and soon Peesapaty and his wife were receiving orders for millions of spoons from customers worldwide.[4]

Today, consumers continue to respond enthusiastically to a product that is promoting sustainable agriculture and reducing the amount of waste in the world's landfills—a win-win situation for the environment and India's millet farmers.

Ironically, plenty of people had been aware of India's groundwater depletion, the ongoing transition from millet to rice cultivation, the excessive use of electricity by farmers, and the wastefulness of plastic utensils. But nobody had focused on the links between the phenomena.

It took Peesapaty some time to connect the cultivation of rice with dwindling groundwater, and at that point, his first response was not to invent edible spoons but to think of new food products to increase demand for millet. It wasn't until his focus shifted to plastic utensils that he began paying fresh attention to the situation. It was only then that he had the inspiration to invent an original solution to a pressing problem.

WHAT IS ATTENTION?

Attention is the active effort of looking at the world to observe problems that need to be solved, opportunities worth addressing, and solutions that can be dramatically improved or revised. Attention has two key attributes: narrowed selectivity—that is, focusing on certain things while withdrawing your focus from other things—and increased cognitive energy—that is, committing more mental energy to the study of the particular thing, person, experience, or context.

Paying attention—devoting more time and resources to studying a phenomenon—improves your chances of noticing interesting and overlooked factors. It also requires that you apply your full concentration to a situation or source of information. This means undivided (not fragmented) effort. Therefore, attention is a selective action. There are so many signals competing for our attention that we must choose where to focus. We must place that focus on some aspects of a situation and largely ignore others. For this reason, especially when we're dealing with large volumes of data, it is easy (and very common) to miss or dismiss weak or ambiguous signals.

For example, if you're a pet owner, you're probably familiar with the "happy dance" (jumping, wagging the tail, purring, and pawing) that hungry dogs and cats perform as their food is prepared. It was only recently, however, that pet food companies recognized what a kick pet owners get out of this behavior. As a result, the companies have designed premium products that require *more* preparation in order to increase the emotional reward for the humans. Health and convenience, it turns out, are not the only criteria that pet owners care about—they just didn't know it.[5]

SELECTIVE FOCUS

When we turn our attention *toward* something, we must necessarily turn it *away* from something else. Because attention is a selective activity, we must choose where and on what to focus. This attention allocation directs how individuals and organizations interact with the external environment. It determines which stimuli they notice and which they overlook.

Narayana Peesapaty wasn't the first person to notice India's declining groundwater levels. But he was the first (that we know of) to focus his attention on the driving forces behind those trends. Having discounted changes in rainfall as the likeliest cause, he turned his attention to changes in farming practices. It was then that he identified the cultivation of rice as a prime factor. At that point, he moved out of his office and started investigating on the ground. He saw the amount of water needed to grow rice; he saw the impact of electricity subsidies on the cost of pumping water; and he saw the surplus rice rotting in warehouses and attracting rats. And he realized that he needed to do something.

THE LIMITS OF ATTENTION

Not all attention results in breakthrough insights. Past experiences prime you for how you look at the world. They influence what you see as important, interesting, or novel, and what you take for granted. Your conditioning, therefore, can interfere with the quality of attention by channeling your focus in particular directions, affecting what you notice and blinding you to radical insights. Simply paying attention to something doesn't always yield breakthrough insights.

In particular, your previous professional experiences may affect what you see. The French call this *déformation professionnelle*—a cognitive bias that causes you to observe reality through the distorting lens of your job, your training, and your profession. Instead of seeing the world as it is, you see it in the way a lawyer, engineer, or graphic designer has been conditioned to see it. This isn't a matter of summoning sufficient will and focus. Your habits, whether professional or cultural, have a powerful influence on your attention. They not only color what you see but also affect the meaning you derive from your observations.

As an example, consider the Finland-based company Stora Enso. At the dawn of the twenty-first century, Stora Enso, founded in 1288, was the world's oldest limited liability corporation and Europe's largest paper and board maker. The company was steeped in history and conventional thinking, and had internalized a particular view of trees that was shared by the senior executive team, all of them steeped in the pulp industry. And therein lay the problem. Without realizing it, this narrow view limited their focus and ability to innovate in a world where sales of paper products were falling fast (thanks to digital publishing) and where the public was increasingly intolerant of timber-harvesting practices that contributed to deforestation and global warming.

In response to declining sales, CEO Jouko Karvinen and his executive team reacted predictably, reducing production and cutting 2,500 jobs. Then Karvinen hired a chemist from DuPont to join the team, and the company's worldview was turned upside down.

According to Karvinen, the newcomer brought with him a very different outlook: "He didn't see the tree in terms of roots, trunk, branches, and leaves. For him, it was cellulose, carbon, and sugar. He said, 'You folks are only using about 45 percent of the value of the tree.' It was a real eye-opener." That simple observation proved pivotal. It encouraged the organization to explore new avenues of business, ultimately prompting centuries-old Stora Enso to reinvent itself as a renewable materials company.[6]

When it comes to generating out-of-the-box insights, the quality of your attention is just as important as the quantity. The freshness and scope of your attention really matters.

ALIENS TO THE RESCUE

Imagine that aliens visited our world from a number of planets. They arrive with different repertoires of knowledge and expertise, but they quickly realize that their backgrounds have little relevance on the new planet they're trying to understand. So they set aside their preconceptions and open their minds to a different reality. They see the world as it is, not as they wish it to be.

Obviously, it can be difficult to empty your mind of all your preconceptions. But you can, at least, articulate your assumptions. Simply listing your initial point of view and expectations helps you become more conscious of your knowledge and beliefs. Once you articulate your frame of reference on paper, so you know what you're fighting against, it's easier to suspend your judgment.

To construct meaning and explore aspects of a situation that you may want to change—in any walk of life—you need to get a richer sense of what's happening, spot the first indicators of change (weak signals), and detect anomalies. In short, you need to find different ways of looking at the world so you can spot things that you or others have missed.

There are two ways to augment the quality of your attention. You can try to see *better*, or you can try to see *differently*.

If the problem is one of achieving more granularity or a wider scope (seeing better), you first need to adjust your zoom. If you're seeking an alternative view (seeing differently), you need to point your lens in another direction and switch your focus. These techniques provide complementary and mutually reinforcing ways of paying attention to different types and sources of information.

ADJUST YOUR FOCUS

To see better, you can either zoom in or zoom out.

Zoom In

You use a close-up lens to observe the situation in more detail, picking up on nuances, incongruities, anomalies, and weak signals. You go deep on a particular aspect of the situation or group of people that seems pertinent or interesting. For example, it was by paying close attention to poor villagers and having deep conversations with a woman selling bamboo stools that Muhammad Yunus came to realize there was a need for a new type of financial help: microcredit.

After obtaining his PhD from Vanderbilt University and teaching in the US, Yunus returned to his home country, Bangladesh, as an associate professor of economics at the University of Chittagong. One day in 1974, he took his students on a field trip to a local village to expose them to the problems afflicting the poor—and the realities of economics. The village would be their university and the inhabitants their professors for the day.[7]

During the trip, his attention was captured by a woman who made beautiful bamboo stools. Through conversations with her, he came to understand the difficulties she faced. She earned just two cents a day because, without working capital, she could not buy her own bamboo. She had to purchase it from a trader who forced her to sell her stools to him at such low prices that she could never put money aside. Caught in a poverty trap, she was unable to create the buffer needed to raise herself above a subsistence level. And she was not the only one in that situation.

Later, Yunus returned with a student to survey the whole village. They spent a week visiting families there and compiled a list of forty-two people who were similarly controlled by moneylenders. They discovered that it would take a total of less than twenty-seven US dollars to extract the sufferers from their predicaments, so Yunus decided to provide them with loans out of his own pocket. "The excitement that was created among the people by this small action got me further involved in it," he recalled."[8] The mini loans were all paid back in full, prompting Yunus to ask, "Why not do more of it?"[9] In this way, the idea for microcredit was born.

Zooming in, either as an observer or participant, helps you avoid being superficial or impressionistic in your observations.

Zoom Out

Here, you use a wide-angle lens to take in larger patterns and trends, and expand the scope of your attention. This is what Peesapaty did when he looked at rainfall trends, shifts in crop patterns, and the impact of energy subsidies on water consumption.

You can also zoom out by engaging different categories of stakeholders with a view on the phenomenon you're trying to change. Zooming out is the mechanism that helps you see the world from multiple angles. For example, a hospital trying to improve the quality of a patient's experience might engage patients, their families, doctors, nurses, insurance companies, start-ups, and other stakeholders to capture equally valuable (and complementary) perspectives on the problems that exist and the solutions that make sense.

Zooming out helps to prevent bias from shaping your thinking by multiplying the sources used in your data collection efforts, allowing you to comprehend more of the total picture. In addition, this lets you spot interesting areas of significance—areas that warrant closer examination.

You need to alternate between zooming in and zooming out to get a clear picture of what is happening—to see the forest *and* the trees. However, the two techniques may require different skills. Zooming in is closer to what an anthropologist does in the field, while zooming out is more like what a sociologist does in studying trends, patterns, and data and separating the properties of a complex system. A practical example of the difference comes from the cofounders of Mission: Launch from Chapter 1. While Teresa Hodge studied the prison system as an inmate (zooming in), her daughter Laurin researched the nationwide figures relating to incarceration, reemployment, recidivism, and collateral consequences on families (zooming out).

SWITCH FOCUS

To see differently, you need to switch focus.

To do this, you often need to redirect your attention to the signals coming from fringe sources—sources you might normally perceive as marginal. Consider the needs of the people on the fringes. Spend time with them.

For example, when US food company Kellogg was looking for new insights on healthier snacks for schoolchildren, they first questioned parents, teachers, nutritionists, and the kids themselves. Although these interviews proved useful, they tended to become redundant, yielding few fresh insights. Who else could they turn to for more original perspectives?

An effort to switch focus revealed additional stakeholders to consult: school janitors. The janitors had tacit knowledge about kids' eating habits, an understanding that could reveal incongruities between common beliefs and the reality on the ground. This turned out to be a rich source of insights because the janitors saw the entire "black market" of school lunches: what was traded and what was thrown away, including a trash can full of apples. (Just because kids take food in the cafeteria line doesn't mean they eat it.[10]) Kellogg leveraged those insights to transform its core snack range.

What you may also find by switching focus is that some people use your products or services in weird ways—ways that can tell you a great deal about the possibilities for innovation. For example, the iconic Danish toy company Lego has plugged in to the internet forums of its adult fans to learn about their design and manufacturing ideas for Lego products.

Since the 1990s, adult Lego users have been pioneering new and more advanced models, and they share their ideas and experiences with one another online. Thanks to their deep knowledge and understanding of the company's product lines, these adult fans have created new strategy games, modular building standards, and even specialized software. By 2012, there were more than 150 Lego user groups on the internet, with over one hundred thousand active adult fans worldwide. Many of these people had been experimenting with Lego products for decades (usually since childhood).

Initially, the reaction of Lego executives and designers to this cornucopia of fan-based innovation ranged from apathy to downright hostility. A deeply private company that tightly controlled its products and intellectual property, Lego even shut down fan forums at one point. Fortunately for everyone involved, the company soon changed its mind, determining that the benefits of collaboration far outweighed the risks.

Since the early 2000s, Lego has formalized its relationships with adult Lego fans through its Ambassador Program, drawing on their enthusiasm, creativity, and expertise to develop new products and marketing programs. The company also solicits feedback and advice about new product lines and designs. Lego's management has learned that these collaborations work best when the fans have areas of expertise that its employees don't—for example, architecture or sensor design and manufacture.[11] These represent niches within the niche.

Lego's online platform, Lego Cuusoo, provides a venue through which users can introduce their innovations to the company. Lego Cuusoo allows enthusiasts to upload their designs to a webpage where other users can vote on them. Models that receive ten thousand votes are reviewed by the Lego Group for their potential for commercialization. If an idea is selected to be commercialized, the Lego Group takes over the development process and the innovator receives 1 percent of the total net sales.[12]

Just as Lego is learning a lot from the innovations of its adult enthusiasts, the cleaning product giant SC Johnson is learning about the shortcomings of its offerings by observing the unmet needs of hygiene-obsessed OCD sufferers, and furniture maker IKEA is picking up new and radical ideas by studying the designs of "IKEA hackers."

As with Lego, IKEA's initial reaction to the creative ideas flowing from its fans' website was to shut down the community. Within twenty-four hours, though, the company realized that legal action was a mistake. Better to join them instead of trying to beat them. Today, IKEA embraces the hacking culture. Its designers participate in hackathon sessions where they tinker with each other's products, and clever hacks seen on the internet (like a working wooden bicycle) are posted on IKEA's office walls.[13]

Switching focus helps you overcome the limitations that arise from focusing too narrowly on the incremental needs of the mainstream population. It also helps you to spot weak signals and potentially wider unmet needs by observing nonstandard users of a service or product. Some users can tell you a lot more than others.

DIGITAL AUGMENTATION

Digital technologies can complement the traditional techniques of switching focus, zooming in, and zooming out, helping ALIEN thinkers pay better attention to the world around them. These technologies offer new ways of noticing tacit needs and enable you to track behavior on a much larger scale without having to observe it firsthand.

For example: In 2010, Nestlé was the target of a negative campaign coordinated by Greenpeace, which posted a video on YouTube condemning the loss of thousands of hectares of orangutan habitat in Indonesia caused by planting palm trees. Though Nestlé was not directly responsible for the habitat loss, as a buyer of palm oil, it was viewed as guilty by association.

Nestlé responded by asking YouTube to take down the video, which it did. By then, however, the video had gone viral. Within a few hours of the video being posted, comments critical of Nestlé began appearing on the company's Facebook page.

Nestlé probably could have defused the situation by responding in a measured manner. However, the people in charge of monitoring Nestlé's Facebook page reacted with frustration and anger at the growing criticism. From the perspective of the critics, the Facebook monitors were acting not as individuals but as the face of the corporation. Soon thereafter, the story became more about Nestlé's poor social media practices than about deforestation.

In the wake of these incidents, Nestlé completely revamped its approach to social media. It established a digital acceleration team (DAT) at its global headquarters in Vevey, Switzerland, consisting of a rotating group of millennials from its offices around the world. Their chief mission was to act as an early-warning system by monitoring social media sites for mentions of the company and its brands, paying special attention to comments that might prove damaging.

As of 2020, the DAT still exists, but the process of social listening has largely been automated through digital tools. These tools, including Radian6, Brandwatch, and Sysomos, are able to monitor online activity on a massive scale and then, if necessary, zoom in on specific events. The automation of social listening allows for "scanning wide"—well beyond the ability of individuals.

Three Benefits of Digital Tools

Digital tools boost the traditional approaches to attention by bringing objectivity at scale, allowing you to zoom in and out concurrently, as well as freeing up your attention to focus elsewhere.

Objectivity

Digital tools can provide more concrete, impartial data to balance out the bias developed through close contact and empathetic understanding.

In health care, for example, researchers are now studying the lived experience of Parkinson's disease by getting volunteers to use their smartphones to measure tremors (thanks to the function that gives you portrait and landscape view), muscle tone (with the microphone indicating the strength of the voice box), involuntary muscle movements (by tapping fast with two fingers on the touch screen), and gait (by putting the phone in your pocket while you walk). They can thus track the efficacy of medication, not just immediately before and after taking it but also in the long term.[14] And what develops is a much richer picture of what respondents actually do, not just what they *say* they do—of the actual behavior of sufferers and their responsiveness to medication.

Scale

Digital tools let you track behavior on a much larger scale than you could firsthand. For example, the German-based company Nivea conducted online analysis (netnography) of discussions around deodorant use across two hundred social media sites. Contrary to expectations, it learned that the key preoccupation of its customers was not fragrance, effectiveness, or irritation but the staining of their clothes. This nugget of information paved the way for a very successful new category of antistain deodorants.[15]

Digital technology also allows you to switch focus and look from another angle. For example, Stanford professor Manu Prakash created Abuzz, a platform designed to combat the spread of mosquito-borne diseases by producing a detailed global map of mosquito distribution. Its purpose is to pinpoint when and where the most dangerous species of mosquitoes are likely to be present, enabling more targeted and efficient control efforts.

Surveillance of mosquitoes traditionally relied on highly trained researchers and scientists capable of identifying the most virulent species. But digital technology has made it possible to engage regular citizens in the surveillance by recording the buzz of a mosquito with their cell phone.

Mosquito species can be identified by the frequency of their wingbeats, which is what produces their characteristic whine. Prakash and his team created a mosquito sound library, organized by species, which powers a matching algorithm. Recognizing that the people they hoped to reach in less-developed regions might not have access to the latest smartphones, the researchers designed the platform to work with recordings made from almost any model of cell phone. In a study, most of the data they focused on was recorded on a twenty-dollar clamshell-style cell phone from 2006.

Contributing to research on where the most virulent mosquitoes may be found is as simple as holding a cell phone microphone near a mosquito, recording its hum as it flies, and uploading the recording to the Abuzz website. The researchers take the raw signal, clean up the audio to reduce background noise, and then run it through an algorithm that matches that particular mosquito buzz with the species most likely to have produced it. As a result, the researchers have been able to pay attention to the mosquito issue in new ways and on a much larger scale. This form of crowdsourcing is low cost, fast, and easy to implement. By switching their attention lenses, the researchers gained an incredible amount of new data about mosquitoes.[16]

Automation

Because digital technologies automate many processes, they free you to focus on the most precious sources of information. Finding people at the margins, who can teach you something new or counterintuitive about a product or service, has always been difficult. Because they operate outside the usual parameters, these people are often hard to find. Traditionally, companies had to go to great lengths to identify and learn from them.

The digital world hasn't solved this problem, but it's significantly eased the burden. In most cases, these unique, weird, or unconventional lead users face the same challenge as the company: they want to connect with and learn from like-minded people. So they self-organize. In many instances, they organize in ways that make them relatively easy to find.

By far the largest forum for these associations is Reddit, an online community of communities that, as of mid-2018, was the fourth most popular website in the United States and eighth most popular in the world. The number of people using the site is impressive: more than three hundred million viewers read 725 million comments per year from almost nine million posters.[17]

Reddit organizes content into communities called subreddits; each subreddit focuses on a particular topic. By July 2020, the number of subreddits surpassed 2.2 million.

Earlier, we discussed how Lego learns from its adult enthusiasts. As it turns out, there is a subreddit specifically for AFOLs (Adult Fans of Lego) with more than twenty thousand subscribers. In addition, there are dozens of other Lego subreddits focusing on different user types, product variations, and play experiences.[18] These communities contain a treasure trove of information and insights for Lego.

Reddit and other online communities provide a shortcut for identifying and paying attention to fringe users, unconventional thinkers, and other interesting stakeholders who are frequently overlooked.

Determining What People Really Care About

With digital tools, you don't necessarily confront the same drain on your time and energy that you do with personal observation. In fact, because digital tools can often blur the boundaries between traditional techniques, they sometimes let you achieve multiple benefits at once. For example, the cases of Nivea and research on Parkinson's disease show how digital tools can extract and deliver richer data on a larger scale (zooming in and zooming out simultaneously). Abuzz is an example of switching focus to involve "citizen scientists," as well as zooming out to see the larger trends. And Reddit allows you to identify niche communities, to zoom in or out, and even to switch focus by following discussion threads in different directions.

Digital technology can help you to gain breadth *and* depth of knowledge by exploring multiple sources of information in parallel rather than in sequence.

The case of Cambridge Analytica is a powerful example of how machines can zoom in on individual patterns of online activities to infer what the personalities of the millions of people involved are like. Cambridge Analytica has received a lot of media coverage, most of which has focused on how the company was able to obtain data on more than eighty million Facebook users and how it allegedly failed to delete this data when told to do so. But there is also the matter of what Cambridge Analytica actually *did* with the data: it devised a new way to zoom in on individuals to generate insights and exert influence.

Pollsters have long used segmentation to target particular groups of voters—for example, categorizing audiences by gender, age, income, education, and family size. Segments can also be created around political affiliation or purchase preferences. But Cambridge Analytica was contracted by the Trump campaign to provide an entirely new weapon for its effort to win the White House. While the company used demographic segments to identify groups of voters, as Clinton's campaign did, Cambridge Analytica also segmented voters using psychographics. Whereas demographics are informational, psychographics are behavioral—a means to segment by personality.

Traditionally, there have been two ways to assess someone's personality. You can get to know them really well—usually over an extended period of time—or you can have them take a personality test and ask them to share the results with you. Neither method is (realistically) open to pollsters. So Cambridge Analytica found a third way. It collected fragments of individuals' online activity and used the data to build personality profiles for millions of people.

One of these approaches used Facebook likes. Whether you choose to like pictures of sunsets, puppies, or people apparently says a lot about your personality. So much, in fact, that on the basis of three hundred likes, Cambridge Analytica's model was able to describe an individual's personality with the same accuracy as a spouse.

Imagine, for example, that you could identify a segment of voters who are high in conscientiousness and neuroticism, and another segment of voters who are high in extroversion but low in openness. Clearly, the people in each segment would respond differently to the same political ad. But on Facebook, they do not need to see the same ad. Instead, each person would see an ad that had been individually tailored to elicit the desired response, whether that response was voting for a particular candidate, not voting for a candidate, or donating funds.

Cambridge Analytica worked hard to develop dozens of ad variations on political themes such as immigration, the economy, and gun rights, all tailored to different personality profiles. They were able to zoom in on individual voters by combining the voter's digital footprint with advanced analytics.[19]

Obviously, the story of Cambridge Analytica is a promise of the level of attention that's possible with new digital tools and techniques *and* a cautionary tale about the dangerous directions in which those possibilities might take us.

Spotting the Signal amid the Noise

Finding hidden patterns in large and varied data sets is hard. Identifying the sources or reasons behind these patterns is even harder. After all, many correlations are spurious.

A large US insurer was very concerned about churn—the frequent policy switching that occurs among customers. Churn is particularly problematic in highly competitive insurance segments such as auto and home, where profit margins are low and customer acquisition costs high. Insurers need to retain customers through multiple renewal cycles to recoup their acquisition investments. When a policyholder calls the company to cancel a service, the call is routed to a specialty contact center that deals with cases of potential churn. The contact center is staffed by trained agents who are supported by sophisticated digital tools that assess the mood of the caller and suggest ways to convince him or her not to leave.

Even with specially trained agents and state-of-the-art tools, however, the success rate in retaining customers was only 16 percent. This meant that 84 percent of customers who called intending to cancel their policies ended up doing so. The company decided that it needed a data-driven approach to improving the retention rate, so it collected data from multiple external sources to augment the sophisticated view of the customer that it already had. This data was then combined and analyzed across tens of thousands of calls to identify patterns that might suggest ways to improve retention. The company knew which calls resulted in success (the policyholder decided to stay) and those that resulted in failure (the policyholder left).

Researchers analyzed the data but couldn't find any obvious patterns that would improve the retention rate. They looked at the traditional factors associated with churn—the number of claims, tenure of service, number of products in the client portfolio, and costs of the policy, along with external factors such as competitor moves and life changes (garnered from social media)—but none of these made much difference to the churn rate or success of retention. So they directed their attention to new sources of data—in particular, data they weren't collecting but to which they had access.

After some effort, they found their answer: the relationship between the caller and agent. When there was an emotional connection between the two, there was a much higher likelihood of the caller staying with the insurer. For example, the researchers found that when a female policyholder in her thirties with children under ten spoke with an agent who was also a woman in her thirties with young children, the likelihood of success (i.e., retention) soared, regardless of all other factors. Previously, the impact of the caller-agent relationship had been overlooked.

This finding was interesting but could add no value unless the company adjusted its process to take advantage of the insight. The standard procedure at the contact center, as at most contact centers, was that an incoming call would go to the first available operator. Callers should not be kept waiting. However, this first-available-operator rule ignored the potential for a connection between the caller and the agent. It assumed that interactions were perfectly fungible—on average, one would be just as good or bad as another.

The key insight—that the relationship between caller and agent was critical to a successful outcome—could not have been found without the insurer's digital data collection and analysis tools. However, even this insight would have been useless unless the company changed its policies, a key navigation challenge (see Chapter 6).

So the company decided to throw out the first-available-operator rule and dynamically assign callers to agents who were compatible, based on the demographic characteristics that were shown to maximize positive outcomes. This was a major change in procedure, one that defied conventional wisdom and was heavily resisted at first. After all, call center managers had been taught throughout their careers that callers should not be forced to wait longer than necessary.

The process of navigation did not stop there. Not only did the insurer change the policy on answering calls, but it also changed its hiring policies so that the demographics of its call center employees would more closely match the demographic profiles of its client base. Moreover, it provided special training to teach agents how to prioritize the aspects of their profiles that were compatible with the callers. The result of these changes was a quick doubling of the customer retention rate, from 16 percent to 32 percent.

Digital technologies will probably not replace traditional techniques. Instead, they allow us to augment and expand the types of insights we generate by providing access to a wealth of unfiltered and unstructured user-generated content. Although you can notice needs in new ways and on a larger scale with digital tools, the human touch will remain important for making sense of those insights—for understanding why people behave in certain ways and what they really care about.

—⊶⊷—

PUTTING ALIEN THINKING TO WORK

NTUC: A Victim of Its Own Success

Organized in the late 1960s, Singapore's NTUC Social Enterprises oversees more than a dozen cooperatives that provide affordable services to low-income citizens—from insurance and health care to food and other daily essentials. By 2016, however, the group was facing an existential crisis. It had succeeded so well in keeping cost-of-living expenses down and lifting the needy out of poverty that it risked losing relevance.

Specifically, the group needed to pay more attention to new social trends, emerging technologies, and Singaporeans' concerns about health care and aging. So the leadership team approached us to help NTUC refocus its innovation efforts and keep delivering value.

Developing Multiple Foci

The fastest way to help executives break free from their usual way of thinking is to expose them to new solutions from different industries or regions. At IMD, we have "discovery expeditions"—curated visits to "reference organizations" designed to help people see things with fresh eyes.

We kicked off NTUC's expedition with a series of online discussions in which the top 120 executives identified common areas of interest. They agreed on four critical issues: cost, collaboration, customer orientation, and new technologies. From there, we developed an itinerary that would address all of these concerns:

- Bangalore, India: Driving efficiency and affordability
- Seoul, South Korea: Innovating in customer engagement
- Tokyo, Japan: Leveraging new technologies for innovation and performance
- Taipei, Taiwan: Fostering collaboration in the organization and the environment

To help the executives rethink their world, we exposed them to a wide variety of organizations spanning finance, education, health care, supermarkets, food courts, and more. For example, the Bangalore trip drew on India's reputation for meeting overlooked demand at the "bottom of the pyramid" and for frugal innovation (*jugaad*). It also leveraged Bangalore's importance as a technology center for international companies and start-ups, including social entrepreneurs. The visits also provided opportunities to put the techniques associated with attention—zooming in, zooming out, and switching focus—in action.

Zooming in: To understand how one company was innovating to better meet user needs, some NTUC executives visited Narayana Health, a hospital chain that has dramatically driven down the costs of cardiac surgeries, putting treatment within reach of thousands of poor Indians. The founder, Dr. Devi Shetty, was described by the *Wall Street Journal* as the "Henry Ford of heart surgery."

Zooming out: To take in the bigger picture, executives visited Akshaya Patra, a nonprofit organization that provides free school meals across India. It combines technology and sophisticated supply chain practices to run state-of-the-art kitchens that serve millions of children. The visit highlighted the interactions between food, health, and education—sectors that belong to separate co-ops within NTUC. In describing Akshaya Patra, its founder and chairman, Madhu Pandit Dasa, said, "It's a hunger-eradication program. It's an education program. It's a social project. It's a nation-building effort." The message resonated deeply with the Singaporean visitors.

Switching focus: To learn from organizations that look at the world from a different vantage point, the visiting executives met with the founders of a number of start-ups, including the online match-making site Wedeterna. This platform respects the Indian tradition of arranged marriages while empowering couples to create their own pro-files rather than leaving it to their parents. Although this was clearly a long way from NTUC's activities, it offered intriguing lessons in how to break with convention and use technology to respond to emerging social trends and user needs.

Seeing New Possibilities

The NTUC executives' exposure to different ideas has paid dividends, generating rich discussions about using technology and disruptive models to meet evolving user needs while remaining true to the group's mission. The expedition inspired the executives to look at their own business in new ways, including a less siloed view of NTUC's activities.

After returning to Singapore, the executives launched bold initia-tives on two fronts.

First, the expedition sparked a flurry of technology-driven inno-vations inside the individual businesses, including the supermarket chain NTUC FairPrice, which overhauled its shopping portal and mobile app to provide a more streamlined shopping experience and a convenient platform for donating groceries to the less fortunate. The expedition also prompted the creation of a new NTUC social enter-prise dubbed MoneyOwl. Launched in November 2018, MoneyOwl offers independent financial advice through a combination of human advisers and robo-technology. The platform uses algorithms to analyze client needs and provide automated financial planning to low-income people who are underserved by banks and conventional financial advisers.

Second, the trips fueled ideas for collaborative innovation. The most powerful example was a pilot program that brought together four NTUC enterprises to facilitate active aging. It started as a discus-sion between executives from NTUC Health, which manages activity centers for seniors, and their colleagues from NTUC First Campus, which runs childcare centers. The breakthrough insight was this: Why not place the senior centers and childcare centers in close proximity? That way, grandparents could help look after grandchildren while the parents were at work. It would also encourage intergenerational bonding, help children learn, and lift the spirits of the elderly. Better still, why not integrate amenities to create a "one-stop shop" with a hawker center, a collection of small food stalls (run by NTUC Foodfare), and a supermarket (run by NTUC FairPrice)? And why not adapt the supermarket to the needs of elderly customers? The changes ultimately included call-assist buttons, magnifying glasses in the aisles to help customers read product labels, and shopping trolleys and customized shelving for easy wheelchair access to products.

Such innovations are revitalizing the group's role as a change agent and thought leader in Singapore—one that refuses to accept the status quo and is ready to experiment, take risks, and collaborate across boundaries.

KEY TAKEAWAYS

- Attention is the active effort of looking at the world to observe problems that need to be solved, opportunities worth addressing, and solutions that can be dramatically improved or revised.
- Attention has two key attributes: narrowed selectivity—that is, focusing on certain things while withdrawing your focus from other things—and increased cognitive energy—that is, committing more mental energy to studying a particular thing, person, experience, or context.
- When dealing with large volumes of data, it is easy (and very common) to miss or dismiss weak or ambiguous signals. For example, it was only recently that pet food companies began designing products to increase the emotional reward for their *human* customers.
- Your conditioning—notably your professional training and work experience—can interfere with the quality of attention by

channeling your focus in particular directions, impacting what you notice and blinding you to radical insights.

- To think like an alien, you must set aside your preconceptions and open your mind to alternative takes on reality. To see the world as it is, not as you wish it to be, you can either zoom in or zoom out.
  - » To zoom in, use a close-up lens to observe the situation in more detail and pick up on nuances, incongruities, anomalies, and weak signals—as Muhammad Yunus did while talking to a woman making bamboo stools.
  - » To zoom out, use a wide-angle lens to take in the larger patterns and trends, and expand the scope of your attention—like Narayana Peesapaty when he tried to establish the causes of groundwater depletion in his region.
- To see differently, you need to switch focus—to redirect your attention to the signals coming from fringe sources, people and things you might normally perceive as marginal. Lego achieved this by plugging in to the internet forums of its adult fans.

QUESTIONS TO ASK YOURSELF

1. Can you recall a time when you missed or dismissed a weak signal that later proved to be important?
2. What biases and preconceptions (personal and professional) could be closing your mind to alternative views of reality?
3. Have you spent enough quality time immersing yourself into the phenomenon you're trying to change? What are the big-picture items and what are the small, interesting details?
4. How can you enrich the way you look at problems that must be resolved? Are there specific categories of people you should spend more time with, places you should visit, dynamics you must observe?
5. Can you expand your search for solutions? Are there interesting "fringe sources" you should investigate or unusual stakeholders you could learn from?

On Digital

1. Can you use digital tools to collect different types of objective data to inform and enrich your understanding—for example, through sensors or mobile phones?
2. Can you use digital tools (such as Reddit) to immerse yourself in niche communities that discuss and/or experience issues central to your inquiry?
3. Can you find new sources of inspiration through an analysis of online discussions?
4. Can you use digital tools to validate the insights you are collecting—for example, by providing objective data?
5. Can you use digital tools to help you zoom in or zoom out—for example, by using digital search strategies?
6. Can you find new sources of inspiration through digital channels?

THREE

Levitation

Elevate Your Thinking

BERTRAND PICCARD IS A MAN OF ACTION. LIKE HIS FATHER (AN ocean explorer) and grandfather (a physicist and hot-air balloonist who inspired the character of Professor Calculus in the Tintin comic books), the Swiss adventurer and psychiatrist is continually in motion.

As a teenager, he catapulted himself from cliffs in hang gliders. By his twenties, he was piloting microlight aircraft. Later, he began flying hot-air balloons, competing against Richard Branson and Steve Fossett to become the first person to complete a nonstop flight around the world in a gas balloon.[1]

But Piccard's greatest achievement took place not in the clouds but on the ground. It happened as he sat motionless in the Egyptian desert, killing time.

THE BIRD'S-EYE VIEW

Piccard twice failed to realize his goal of a nonstop hot-air balloon flight around the globe, but on a third attempt, in March 1999, he and his copilot, Brian Jones, took twenty days to float to the finish line near Dakhla Oasis, Egypt, aboard the custom-built *Breitling Orbiter 3*.

During the five-hundred-hour trek, the balloon was kept aloft by a mixture of propane and helium. Although it carried 3,700 kilograms of the liquefied gas, housed in long silver cylinders suspended from the craft, Piccard was continuously stressed and obsessed with the fuel situation—for good reason. For long stretches, they were far beyond sight of land or the reach of would-be rescuers. By the time the men touched terra firma again, in a remote stretch of desert 220 miles from the Nile, only forty kilos of gas were left. They'd barely made it. As he sat in the sand, Piccard resolved to never let that happen again.

But how?

The only variable that Piccard could control with the *Breitling Orbiter* was altitude (by adjusting the gas to catch the best airstreams above or below the craft). Lugging along more fuel wasn't practical, and he had no realistic options for conserving fuel while in flight.

At the end of the trip, Piccard found himself in an unusual situation. Until he and Jones were picked up and driven back to civilization, they had nothing to do—and ample time in which to do it. "I was sitting in the Egyptian desert, my back resting against the capsule of my hot air balloon, gazing at the horizon. The wind, which had pushed Brian Jones and I during 20 days non stop around the world on board Breitling Orbiter 3, only just allowed us to accomplish our round-the-world flight, despite the fear of seeing our fuel supplies run out. I couldn't stop now. This success had to be a means to an end, not an end in itself."[2]

This six-hour period of forced inactivity proved critical. They had no distractions and plenty of time to reflect.

Then it came to him: the core problem was not how to manage fuel but how to manage *without* fuel—using only renewable energy. He would fly around the world again, this time with no fuel at all. This would solve his immediate problem and also serve as a bold statement against our dependence on fossil fuels. "I had to show that exploration can go from the discovery of new continents to the promotion of a better quality of life."[3]

THE BIRTH OF *SOLAR IMPULSE*

With this new understanding, Piccard came up with the idea for *Solar Impulse*, a solar-powered craft capable of perpetual flight.

Piccard was flush with excitement, but he had no illusions about the difficulty of the tasks ahead. In the early 2000s, when he began earnestly searching for allies and business partners, the clean energy technologies needed to power his vision—solar cells, ultralight aircraft, large-capacity batteries, and electric motors—were well established but not widely used. In addition, they were somewhat primitive compared with the technologies developed since then. At that time, for example, several solar airplanes (manned and unmanned) had been built, but none had flown for days with a pilot on board, and none had crossed the earth's vast oceans. Worse: a solar plane had recently broken up and plunged into the waters off Hawaii after hitting strong winds.

In 2003, Piccard presented his vision of a solar plane to the Swiss Federal Institute of Technology, which agreed to a feasibility study. The study was led by André Borschberg, an engineer, businessman, and former fighter pilot in the Swiss air reserves. His conclusion? Although the concept of a solar-powered aircraft was possible, it was beyond the reach of existing technology.

This verdict might have disappointed most people, but it didn't faze Piccard—or Borschberg. In fact, it energized them. As soon as the feasibility study was done, Borschberg signed on with Piccard to found the Solar Impulse project. Not long afterward, the two began designing a plane from scratch.

Thirteen years and twenty-four thousand miles later, their brainchild—the *Solar Impulse 2*—touched down in Abu Dhabi, having traveled around the planet without using a drop of fuel.

The plane was powered by more than seventeen thousand photovoltaic cells, each about as thick as a human hair. They generated enough power for the four electric motors, but the plane wasn't about to break any speed records. Its average velocity was about forty-three miles (sixty-nine kilometers) per hour, and the craft was so fragile that it had to avoid strong winds. In addition, the plane could accommodate only one person at a time (the pilot) in a very cramped space. In short, solar planes aren't about to replace commercial jetliners any time soon, but *Solar Impulse 2* did demonstrate that solar-powered aircraft can work—and if they're not commercially viable today, then they will be in the near future. And the implications of that are profound.

"Climate change is always presented as a battle to protect nature against business and comfort," said Piccard. "Ecologists have put nature before humankind, and that's a big mistake, a false equation. These technologies are available now. We *can* change the way the world works, not by making people's lives smaller, but by making them bigger."[4]

WHAT IS LEVITATION?

In the thick of the action, your mind is filled with confusing information. You're overwhelmed. You have difficulty distinguishing the noise from the signal. To make sense of what you're seeing, thinking, creating, or doing, therefore, you must disengage from your regular activity. You must decenter. You must *levitate*.

Levitation is the act of stepping back to regain perspective. It's critical to creativity in any profession. It's about distancing yourself from the activity itself to think more clearly. To contrast it with attention, levitation is what happens when you stop noticing and cease to look for input by either zooming out or zooming in.

Levitation is not outward focused. It's inward focused and introspective. Your aim is no longer to sense the environment but to *make sense* of it, to let your observations sink in and work out what they mean or could mean. It's about framing or reframing the problem and avoiding the rush to action.

Underinvesting in levitation can lead you to formulate a problem or opportunity incorrectly or to ask yourself the wrong questions. This won't stop you generating good ideas, but you may be heading toward a dead end. Solving the wrong problem has actually been ranked as the number one reason for innovation failure, accounting for over a quarter of all flops.⁵ In fact, the risk of jumping mindlessly into problem-solving mode is heightened by advances in artificial intelligence. As one industry observer famously put it: "The formula for the next 10,000 start-ups is that you take something and you add AI to it."⁶

When you levitate, you set aside your default responses to think twice about what you're experiencing—to expand your thinking and prepare for what should come next.

Piccard was able to pull back and think differently only after he stepped off the activity treadmill. Once his mind was no longer preoccupied, he was able to train his thoughts on a new idea: renewable energy. By redefining the problem, he found the solution and set the stage for his next circumnavigation challenge—this time in a solar-powered plane.

Similarly, Bakeys's founder Narayana Peesapaty was only able to identify the crux of the problem and lay the mental foundation for a solution after stepping back to make sense of India's agricultural situation. Once he was able to gain perspective, he realized that since market forces had created the problem, they could also solve it. He reframed the challenge from one with a public policy solution to one with a market-based solution that would help revive the demand for millet.

You can't prime yourself for inspiration while you're in the thick of the action. To process what you've observed and learned, you need to detach yourself. You need to take a strategic break.

MA THINKING

Levitation may seem like an (ahem) alien concept, but its roots stretch back to antiquity. Self-reflection is at the heart of most philosophies and most ways of living, regardless of culture.

In classical Greece, for example, Plato compared ideas to birds flying around the aviary that is our brain. Until the birds settle, it's difficult to distinguish those representing true knowledge from those representing false knowledge. But in order for the birds to settle, they need a stable perch—a calm and reflective mind.

In Japan, the concept of *ma* (roughly meaning the space or time in between) is rooted in the ancient Shinto religion. *Ma* is considered a driver of creativity, not just in activities such as architecture, design, and the arts, but also in business.[7] Literally, *ma* means the light shining through the gap around a door when the door is shut. It's through gaps and openings that new phenomena and events emerge. *Ma* is also the interval when thoughts mature before becoming full-fledged ideas. Whereas a Westerner might refer to the space between a table and chair as "empty," the Japanese would refer to the space as "full of nothing."

Similarly, most forms of Buddhist meditation are designed to help the practitioner achieve a state of pure awareness by quieting the conscious brain—the noisy, ever-jabbering "monkey mind" that produces a ceaseless stream of conversations, thoughts, narratives, and attempts at labeling everything perceived by the five senses. By contrast, pure awareness is "empty." It's an infinite space of emptiness filled with infinite possibilities. When a mind is completely filled with something, there's no room for anything else. When it's empty, there are no limitations.

The case of Marcus Raichle is a good illustration of *ma* thinking. In the mid-1990s, Raichle, a neuroscientist, began compiling a folder of curious brain scans, which he labeled "MMPA" (for "medial mystery parietal area") and then filed away. It took him several years to collect his thoughts and make sense of the mystery data, but when he did, the resulting breakthrough was a paradigm-shifting moment in the history of neurology.[8]

In the early 1990s, access to new imaging technology fueled a wave of experiments focused on memory, language, perception, and attention. Basically, researchers began putting people in scanners and watching their brains at work. Test subjects were asked to perform particular tasks—such as generating words or judging the movement of a dot array—while undergoing a scan such as an fMRI.[9] In response to the task, certain areas of their brains would light up. These results were then compared to the brain patterns of a control group (test subjects at rest). Because the control group was unstimulated, researchers assumed that their brains were passive.

But Raichle (and others) observed that while certain areas of the brain were lighting up as subjects performed the different tasks, other areas were quieting down. For years, these inconvenient observations were dismissed as "noise." They were frequently disregarded or, in some cases, not reported because they were not the focus of the experiments. There was no place in the researchers' thinking for this "in-between" data.

It was several years before Raichle realized that the original focus of the studies was actually a sideshow. The real action was the incredible brain activity that occurred when the subjects *stopped* doing stuff. In the absence of externally driven tasks, the subjects were letting their minds wander, thereby activating an elaborate collection of disparate brain regions that Raichle dubbed the "default mode network" (DMN). This mental network lights up during the freewheeling state that was formerly called "the mind at rest."

It has since been discovered that the "mind at rest" is far from quiet. In fact, the DMN is often more active than a mind that is performing a task. The energy consumed by the DMN is about twenty times that used by the brain when consciously responding to outside stimuli.[10] In addition, it's now known that the DMN is involved in (among other things) introspection, remembering the past, envisioning the future, and understanding others' thoughts.

This case is a classic example of what writer Steven Johnson calls a "slow hunch." This innovation process often starts with an anomaly that points to a larger truth, but it can take years of contemplation for the hunch to crystallize and come into focus as a tangible idea.[11]

WHY LEVITATION MATTERS

Levitation helps us overcome framing and action biases, prompting us to do any or all of the following:

- Question our initial assumptions
- Redefine the problems we want to solve
- Uncover new insights
- Reflect on what's really important (and what's not)
- Distinguish important—but sometimes weak—signals from noise

Because Teresa Hodge took time to levitate—to try to understand the situations of her fellow inmates and their families—she was able to make sense of why so many women kept coming back to prison. Eventually, levitation helped her locate the answer: because they were denied opportunities to obtain good jobs, loans, and affordable housing. Recalling the process itself, Hodge said, "It was as if prison provided a moment of clarity—a place to plan and figure out life and to determine how to start over."[12]

The ability to focus on a task or problem at hand is normally considered a positive trait. But too much of a good thing can have negative repercussions, and it turns out that excessive focus is more likely to inhibit, rather than accelerate, creativity and innovation. Excessive focus actually makes us less alert to the possibilities and opportunities at the intersections—in the *ma* spaces between things. For example, although Procter & Gamble's Gillette group had a toothbrush unit (Oral B), an appliance unit (Braun), and a battery unit (Duracell), it was slower than its competitors to develop a battery-powered toothbrush. Each unit was too focused on its own products and innovations to make that leap.[13]

Paradoxically, switching from externally directed attention to internally directed reflection empowers both individuals and entire organizations to think and move forward more creatively. Levitation is also vital when deciding how to alter a creative strategy and determine which of the other ALIEN thinker strategies to pursue next.

TAKING TIME

Most people, especially those in organizations, have a sense of urgency because they feel that time is short. But ALIEN thinkers are not as beholden to time constraints as most people. To continually innovate, they make time for pauses. They bend time to their will by leveraging two forms of levitation: time-outs and time off.

Time-Outs

As in sports, the purpose of the time-out is to take a step back from the action and chaos in order to reflect consciously on your approach and how best to redirect your efforts. You're going off-line to pursue your thinking under calmer circumstances.

The time-out has two components: figuring out what you think and reviewing how you think.

Figuring Out What You Think

You need to pause to fully absorb what you've discerned and to work out what it really means. Scholars call this *sensemaking*. It is exemplified by psychologist Karl Weick's famous question: "How can I know what I think until I see what I say?" Even for the lone innovator, it often helps to take the time to explain, even to someone unfamiliar with the problem, why you are stuck—or to write it down or say it aloud to yourself.

Sensemaking is concerned with the construction of new meanings, creating order from confusion and chaos.[14] It's often necessary to reframe the question in order to pay better attention.

Sensemaking may be triggered once the individual or group notices events, issues, or actions that are somehow surprising or confusing— discrepancies between expectations and reality. Such anomalies are sometimes experienced as a palpable irritation, uneasiness, or unsettledness.[15] For example, while in prison Teresa Hodge was deeply frustrated to see releasees returning—especially women who had walked out of prison determined to change their lives. She felt compelled to figure out what was going on.

Sensemaking plays a vital role in the learning process that underlies innovation. It can force individuals and organizations to confront discrepant cues—oddities and anomalies that may otherwise be ignored, accommodated, explained away, or normalized.

Talking with her fellow inmates and rebracketing the cues enabled Teresa Hodge to realize that most of them had something in common—that their high recidivism rates were not an anomaly but a logical result of a system that continued to punish offenders long after they were released.

Sensemaking is facilitated when you have slack time between scheduled commitments. If that doesn't happen automatically because you're too busy, you need to block out time to step back. The process can be individual or social. It sometimes takes place in individual heads and sometimes through discussion with others.[16] Sensemaking is needed to develop new systems that incorporate more of the data discrepancies so that the narrative you develop—for yourself and potential stakeholders—is more comprehensive and resilient in the face of criticism.

Let's return to Marcus Raichle. For years, researchers discounted the brain activity of subjects at rest as noise. As one researcher put it: "They fell victim to the always present bias to ignore what is least understood."[17] Only with reflection did Raichle realize he had been focusing on the wrong findings all along. Everyone was interested in how the active brain works, not how it rests. As a result, they had missed the big picture. The more interesting question was the brain's unexplained activity while "at rest"—in the control condition.

To achieve breakthrough solutions, you have to step back periodically to revitalize your understanding.

Reviewing How You Think

Another reason to pause is to engage in what scholars call *metacognition*. This is the process of thinking critically and reflectively about your own thoughts—not the issue itself but your approach to the issue.

This is exactly what Bertrand Piccard did. Upon landing his balloon in the desert, he took advantage of the time to question his own thinking. As a practicing psychiatrist, he asked himself why he'd been so anxious throughout the flight. The answer: it was not so much fear for his own safety as the fear of not completing the trip because of a lack of fuel. This introspection led Piccard to the insight that to avoid the source of his anxiety, he needed to fly without fuel. Instead of thinking of it as a fuel problem, he reframed it as a challenge of dispensing with fuel entirely.

As part of the innovation process, you periodically need to stop doing what you're doing and actively consider not only what you've learned but how you're approaching the challenge. Pausing gives you the opportunity to review your creative process: what you're doing and why. In addition, levitation gives you the capacity to reflect on whether you need to implement any of the other four ALIEN thinker strategies. Take attention, for example.

Reflecting on Your Attention Strategy

As we saw in the previous chapter, to see what others don't, you must find different ways to look at the world. This includes where you focus your attention, whom you target, and how you interact with them. To develop an effective attention strategy, you need to consider the following questions:

- **Where** are you focusing your attention? Are you looking where others aren't? Is your scope broad or narrow? On the one hand, you want to eliminate noise by having a narrow focus that shields you from distractions. On the other hand, you want to maintain a certain flexibility to notice seemingly irrelevant sensory information. There is evidence that real-world creative achievers have broad or "leaky" attention, which may pave the way for divergent thinking.[18]
- **Why** are you looking in this direction? Are you driven by a threat or an opportunity, by aversion or attraction? Are you problem-solving or problem-seeking? Depending on whether you face an unsustainable situation or are proactively trying to identify unmet

needs, the answer to this question will affect the intensity and urgency of your efforts.

- **What** is the object of your inquiry? The way you formulate the question(s) you want to answer shapes the nature of your attention. Also, what do you hope to find: an answer or an insight?

- **Whom** are you paying attention to? Who might offer fresh perspectives? Beyond the specific people who will be using your solutions, there is a whole ecosystem of stakeholders who have fresh views to bring to the issues you care about.

- **How** are you paying attention? Are you *watching* what people do, or *doing* what they do (fully immersed)? Or are you looking at data that reveals what they actually do, as opposed to what they *say* they do? Also, attention can be top-down or bottom-up.[19] In other words, we can actively direct our attention to aspects of the situation that are expected to be important, or we can be more open to surprises, allowing the most striking aspects of the environment to capture our attention.[20]

- **When** are you most alert? Attention can be transient or sustained.[21] It can be episodic (as when you receive smartphone notifications) or continuous (as when you register all the novelties in an unfamiliar country). In the earlier pet food example, the critical observation moment was the interaction between pet owners and pets *just prior* to eating.

This shows how levitation might improve your attention strategy, but a similar approach could help you reflect on and reframe your imagination, experimentation, or navigation strategies. The same six prompts—where, why, what, who, how, and when—can be adapted to stimulate your thinking in new directions on these other dimensions of the ALIEN framework.

ALIEN thinkers are willing to question their observations and conclusions, to critically assess the frameworks they are using and discard ineffective ones. The spirit of levitation is to contemplate. But to do that, you need to pause.

Time Off

Creativity can benefit from conscious analysis of the situation, and also from unconscious thinking. To be more creative, sometimes you need to disconnect completely—to take a vacation from your current preoccupations and tap the power of the unfocused brain. Downtime lets you regenerate your mind and engage in freewheeling thinking.

As an example, consider the pioneering chef Ferran Adrià, whose culinary techniques meld haute cuisine, art, and science. His approach helped him invent more than 1,800 signature dishes over twenty years, earning his restaurant, El Bulli, the rating of "world's best" five times according to a poll of industry experts in *Restaurant* magazine, a record.

The key to his creativity? Closing his restaurant for six months of the year.

"The pressure to serve every day doesn't offer the kind of tranquility necessary to create as we would like," he said. "The most important thing is to leave time for regeneration."[22]

Levitating gives Adrià the ability to invent by rethinking his approach to attention, imagination, experimentation, and navigating the ecosystem. It's the fuel that allows him—and you—to energetically pursue every ALIEN thinker strategy in an enriched way.

"It's important to 'oxygenate' ourselves a bit," said Adrià, "to let ourselves recycle and to adapt our vital and mental rhythms to a new set of demands."[23]

Another proponent of the creative sabbatical is Austrian graphics designer Stefan Sagmeister. In a 2009 TED Talk, he explained that he initially tried to leave his sabbaticals unplanned, assuming that "the vacuum of time would be wonderful and enticing for idea generation."[24] But this approach turned out to be "rather disastrous": Sagmeister mostly just reacted to events or trivial requests. So he started thinking about what he was most interested in discovering during his sabbaticals and prioritizing those experiences. In short, he adopted a much more structured approach to restoring his mental resources.

Sadly, most of us are unable to take a sabbatical abroad. Fortunately, there are less exotic and time-consuming ways to replenish your mind. For example, Kevin Cashman, the global leader of CEO and executive development at the consulting firm Korn Ferry, proposed stimulating your thinking by visiting museums, which he called the "custodians of epiphanies"—that is, the collectors *and* stimulators of inspiration. An even more practical way to recharge, while at work, is to simply switch tasks or projects. A change or distraction may provide just the break you need to disengage your fixation on an ineffective approach.

Taking Mental Breaks

From everyday experience, you know that disconnecting from what you're doing can help you find a path forward. A break as short as five minutes has been shown to help individuals contemplate competing options and make better decisions.[25] At times, the answer to a stubborn problem will pop into your head after a restful sleep or a relaxing walk.[26]

Kevin Cashman's research, which is supported by our own findings from surveys of international executives, confirms this classic experience.[27] According to Cashman, 78 percent of executives say their best ideas come in the shower, while exercising, or while driving or commuting to work.[28]

What do these activities have in common? They are mundane tasks that we tend to perform on autopilot. Sometimes, a moment of lucidity or insight only arrives after you stop thinking directly about a problem. Thanks to the breakthrough DMN research described earlier, we now better understand why. When you engage in activities that you can perform absentmindedly, your brain activity resembles that of the brain in default mode.[29]

One study conducted at the University of Cambridge sought to discover whether the DMN might help us do things—for example, tying our shoelaces or driving along a familiar road—without paying much attention.

To investigate, the researchers asked twenty-eight volunteers to learn a novelty card game while lying in an fMRI brain scanner. Each person was presented with four cards. They were then given a fifth card and asked to match it to one of the other four. Participants were not told the rules—they didn't know whether to match the cards by color or shape—but through trial and error, each person figured things out after a few rounds.

While they were mastering the game, their brain activity resembled patterns that are typical of learning minds. But once participants knew how to match the cards without much thinking, their brain activity resembled people using the DMN, and their responses became faster and more accurate.

This suggests that when we "switch off," our brains go into an autopilot mode that lets us perform tasks reasonably well without thinking much about them. This might also help explain why some tasks, such as playing a well-known tune on a musical instrument, suddenly seem much more difficult when you go from doing them absentmindedly to consciously thinking about them.

The DMN (also known as the "daydreaming network") has been shown to facilitate imagination, creativity, thinking about the future, and reflection.[30] When you're busy, that activity is suppressed, as reflected in an observation popularly ascribed to Einstein: "Creativity is the residue of time wasted."

Paradoxically, lack of attention feeds creativity. Research shows that students with ADHD score significantly higher on tests that measure creativity, due mostly to their difficulty staying focused.[31] According to the lead author of one study, "The very mechanism that allows you to focus can also keep people in a box, and in order to break out of that fixed thinking you have to allow some chaos into the mind."[32]

So when you get stuck, more concentration is not necessarily the answer. You sometimes need to step back to elevate your thinking. How can you do this without overhauling your daily routine or taking a shower?

One simple way is to take a walk. Some of the most fertile minds in history, including philosophers and writers such as Kierkegaard, Thoreau, and Dickens, regarded walks as rituals sacred to their creative routines. Dickens, perhaps the greatest of the Victorian novelists, was a man of strict routine. Every day, he would write from 9:00 a.m. to 2:00 p.m. Afterward, he would put his work away and go for a long walk. Sometimes he walked as far as thirty miles; sometimes he walked into the night. "If I couldn't walk fast and far, I should just explode and perish," he wrote.[33]

No time for a thirty-mile hike today? As an alternative, especially if you're trapped in the office, stare out the window. We often do this instinctively when we're puzzled. It's not to discover what's happening outside but to discover what's happening in the recesses of our own minds. Activities such as window gazing don't guarantee insight, but taking this sort of mental break can help you snap out of an intellectual impasse and regain a sense of agency.[34]

Daniel Pink, author of *When*, recommended that you "take 10 minutes. Go out for a walk without your phone. We're talking about those kinds of breaks. . . . And one of the things that I've discovered, and in fact changed my own behavior on, is that my view always was: amateurs take breaks; professionals don't. And it's the exact opposite. Professionals take breaks. Amateurs don't. Breaks are part of performance. They're not a deviation from performance."[35]

Anders Ericsson (the scholar behind the ten-thousand-hour principle made famous by Malcolm Gladwell) was also a passionate advocate for breaks and recuperative naps. Renowned for his research on expertise, he found that a common habit among elite performers, from musicians and athletes to chess players and scientists, was to work in ninety-minute cycles and then pause to renew energy.[36]

"Take more and better breaks," Pink said.

Also, treat breaks with greater respect. In my view, the science of breaks is where the science of sleep was a decade ago—about to break through the surface. . . . Allowing people a choice, or a measure of autonomy, is critically important. We know that breaks when we move—say, taking a walk—are more restorative than those when we're not moving. Breaks in nature—simply going outside—are especially powerful. And there's a bunch of research that shows that breaks are more effective when we're *fully* detached—not semi-detached. So leave your phone on your desk when you take a break for a walk around the block. Also—and this is key—something is better than nothing. Breaks need not be long to be restorative.[37]

It turns out that downtime not only helps recharge your attention batteries but also encourages unconscious mental activity. The resting brain is far from purposeless or unproductive.[38]

USING TECH MINDFULLY

Levitation is especially important in the digital age, thanks to the daily onslaught of virtual distractions. Although digital tools can facilitate original thinking, they can also constrain it.

Once upon a time, there was a delay, known as a "float," between when information emerged and when it was widely available. This delay gave you time to reflect before you had to act on the new information. Today, the float has disappeared. If something happens, you hear about it instantly, and as soon as you do, you're expected to respond. Even when you're trying to innovate, there is a sense of urgency—frenzies and fast iterations and pivots—to find a product-market fit. In that context, having more time to think becomes a competitive advantage.

Digital tools can help increase your mental processing capacity. By liberating your time, technology lets you focus on value-adding thinking. Browsing the internet and social media can also function as time off, giving you a brief respite from relentless cognitive work.

But beware of false levitation.

False Levitation

False levitation comes in two forms: digital saturation and the echo chamber.

Digital Saturation

The biggest impediment to levitation is never fully disconnecting, and digital tools often prevent a complete disconnection. For example, perhaps you take a walk to transition from reactive mode to sense-making mode, but you take your phone with you. Or you meet off-site to engage in deep thinking about your approach or dynamics, but your eyes keep darting to your mobile device, where emails and calls are piling up. Or perhaps you tend to turn to technology whenever there is a lull. In doing so, you're preventing your subconscious from engaging. You sometimes have to "protect your boredom," as science writer Jonah Lehrer put it.[39]

It takes a conscious effort to resist the addictive pull of technological distractions that instantly fill your spare moments—and all the more so with social media. Engaging with Twitter, Instagram, or Facebook triggers a release of dopamine, similar to eating or having sex. (One study revealed that many eighteen-to-thirty-five-year-olds think abstaining from social media would be harder than not drinking, smoking, or sleeping.[40]) Instead of clearing your mind to wander, you're filling it up with random thoughts. You remain in consumption mode, snacking on other people's thoughts instead of forcing yourself to think.

The Echo Chamber Effect

Technologies can also create an "echo chamber" effect that buffers you from novel sources of ideas. Because the notifications you opt to receive are tailored to your interests, you discard the creative freedom that comes with browsing unless you make a conscious effort to seek out alternative perspectives.[41]

Sometimes, levitation *appears* to occur but is actually compromised by algorithms that are programmed to comply with our preexisting perspectives (think Amazon, Netflix, or YouTube). It's also compromised by social media networks that reflect the recommendations of like-minded thinkers instead of potential disrupters or provocateurs. The echo chamber is so common that you really need levitation to avoid it. The challenge is using the internet to expand your horizons, not reinforce your existing interests, preconceptions, and prejudices.

Levitation Unplugged

When you take a time-out or time off, you sometimes need to unplug. Perhaps one reason that breakthrough ideas often come to us in the shower is because it's one of the last remaining holdouts from the reach of digital tech. It's a haven where you can protect the quality of your levitation, where downtime isn't systematically drowned in distractions. Fortunately, there are also a few other ways to foster levitation.

Block Distractions

At an individual level, you can leverage opportunities to levitate. For example, when traveling to work, switch off the radio and the phone, and let your mind wander. Carve out an hour that's uninterrupted by technology.[42] At team retreats, have people deposit their mobile phones into a basket for the duration of the session.

Fight Tech with Tech

Ironically, you can also harness technology to help you avoid technology. For example, an app called Moment helps you monitor your phone usage—not just how much time you spend on your phone but also how many times you pick it up "just to check."[43] There are also hundreds of mindfulness apps designed to help you step back and regain perspective.[44] Mindfulness sessions as short as ten minutes have been linked with greater creativity.[45]

ALIEN thinkers are conscious of how they levitate because they know that levitation recharges their creative batteries, enhances their ability to make sense of seeming chaos by providing much-needed perspective, promotes divergent thinking, and helps them reframe their approach to the other ALIEN thinking strategies.

+‧═◇═‧+

PUTTING ALIEN THINKING TO WORK
Axel Springer: Too Busy to Reflect

In 2012, German media giant Axel Springer was six years into its largest-ever corporate transformation journey. In 2006, CEO Mathias Dopfner had announced a new company ambition for digital sources to provide 50 percent of total revenues and profits within ten years. At the time, digital revenues were in the single digits and profits were negligible.

In the first years of the transformation, the company shed many of its core assets in traditional print media and reinvested the proceeds in new areas such as job boards, real estate sites, and digital start-ups. The transformation was going well, but the bulk of the remaining core business, like media properties *Bild* and *Die Welt*, was proving difficult to change. Over several years, our close involvement with Axel Springer helped us gain valuable insights into different forms of levitation and how to stimulate it by breaking with normal routines and comforts.

A Change of Scenery

Dopfner felt that the transformation needed to come from the top, and though senior management was supportive, it was proving harder than expected to push their thinking outside normal ranges. So he decided to take a radical approach and sent three of the company's most senior executives to Silicon Valley for nine months. Nothing like this had been done before, and many within the company questioned the move. What would they do there? Who would run their businesses while they were gone? How could the expense be justified?

Despite internal and external questions and criticism, the editor-in-chief and publisher of *Bild*, the chief marketing officer of the media division, and the CEO of one of the business units left for Silicon Valley in September 2012. The goal was to open up their thinking by networking with companies and universities in the area. The results were impressive. All three executives came back energized about new ways to build and adapt the Axel Springer organization.

One of the executives, Kai Deikmann, the publisher and editor-in-chief of *Bild*, left as a quintessential German executive in a custom-tailored suit and returned to his Berlin newsroom sporting a hoodie, sneakers, and a lumberjack beard. He also came back with a clear message to his staff: "I told them we must be ready to make mistakes," he said, "and to see that failure can be a precondition of success."[46]

The value of the Silicon Valley sabbaticals was deemed so high that Dopfner made it a permanent feature. "What began as an experiment, limited in time and personnel with only three employees," he explained, "has now become a regular visiting program. The personal contact to Silicon Valley and the proximity to developments for future digital businesses has proved very valuable for Axel Springer. We now want to build on this and provide further employees with the unique opportunity and challenge to focus on new developments for a certain period of time."[47]

Each visiting fellow spends three to six months in Palo Alto, California. They either put their regular duties on hold for the duration of their absence, turn over these duties to another employee, or, if possible and expedient, continue to perform their duties while away. Any Axel Springer employee who plays a key role in the digital transformation of the company can apply for acceptance to the program.

Breaking Habits and Routines

Although the visiting fellow program provided dividends, it could not address the speed or scale of change that was required. Most people in the traditional business were still resistant to new ways of thinking. To address this challenge, Dopfner and his HR partner, Alexander Schmid-Lossberg, decided to take the top seventy executives on a learning expedition. Often, such collective time-outs take place off-site. However, even with minimal distractions, participants can easily fall into habitual and unreflective ways of thinking and interacting. To combat the pull of preset ways of thinking and encourage shared ALIEN thinking, the pair took a different route. They concocted a trip designed to help the company disrupt itself. The theme of the expedition was "outside your comfort zone." They chose this name because they felt that the participants would only challenge their thinking if they were placed in new and sometimes uncomfortable positions.

First, they flew the senior executives, in economy class, to Silicon Valley for three days. Standing over two meters in height, Dopfner found it a particularly long flight! Second, they stayed in a crummy hotel in a dodgy area of San Francisco and even shared rooms and king-size beds. This was meant to break their habits, help them regain perspective, and make them more receptive to the novelty they encountered in visits to the digital giants and start-ups.

"That place was kind of funky," declared Dopfner, "I think, a little rock and roll, definitely not what we were used to and perhaps made a few of us be more in tune, more sensitized to, more attentive to, more excited about, what's going on around us. And that alone is half the battle. Plus, all the other little things, like sharing double rooms and even beds. By doing this, we hoped it would lead us to having conversations that we wouldn't have in a single room or on the phone. And that was indeed the case."[48]

Schmid-Lossberg concurred. "If you are a coming from a traditional business, say a printing facility, you may ask yourself in the beginning, 'Why should I go there? Does it affect my business?' The head of finance might ask the same questions. But, at the end, everyone took home something. One could see how some of these companies, like Apple, like Google, like Airbnb, work, what they do differently, what could be adapted for our purpose. One could likewise see what should never be adapted. Because not everything in Silicon Valley can be adapted to the European culture, and we don't want to adapt everything. It helped . . . open us up to what is shifting and adapting."[49]

Ready to Reinvent

The combination of time off and time-outs employed by Axel Springer helped them reinvent their entire strategy, resulting in a big strategic shift from content company to platform company. Between 2012 and 2016, the company invested in a series of digital platform companies and, in its biggest acquisition, paid $343 million for a controlling stake in Business Insider, the New York–based financial news site. By 2016, Dopfner's ten-year ambition had been achieved: 60 percent of revenues and profits came from digital sources.[50]

⊹⊱──⊰⊹

KEY TAKEAWAYS

- Levitation is the act of stepping back to regain perspective. It's about distancing yourself from the activity itself to think more clearly. Levitation is not outward focused but inward focused and introspective.
- Levitation may seem like a new and alien concept, but its roots stretch back to antiquity. In Japan, for example, the concept of *ma* (roughly meaning the space or time in between) is rooted in the ancient Shinto religion and is considered a driver of creativity.
- Science now recognizes that a mind "at rest" is far from quiet. In fact, the default mode network (DMN) is often *more* active than a mind that's performing a task.
- Levitation helps you overcome framing and action biases, prompting you to question your initial assumptions, redefine the problem you want to solve, uncover new insights, reflect on what really matters, and distinguish noise from weak signals.
- Because Teresa Hodge took time to levitate—to try to understand the situations of her fellow inmates and their families—she was able to make sense of why so many women kept coming back to prison.
- To innovate like an ALIEN thinker, you must leverage two forms of levitation:
  - » Time-outs: stepping back from the action and confusion to reflect consciously on your approach and how to redirect your efforts
  - » Time off: taking a mental or physical vacation from your current preoccupations to tap the power of the unfocused brain

QUESTIONS TO ASK YOURSELF

1. Can you recall an instance where stepping back from the action led to a sudden insight—a major or minor eureka moment?
2. Can you make time for levitation in your weekly routine? Are there natural moments of reflection—such as during your commute—that you could more systematically leverage?
3. Could you transform wasted time (e.g., sitting at the doctor's office, queueing at the supermarket, boarding your plane) into opportunities for introspection and learning?
4. Should you try to switch tasks, projects, or settings to avoid getting trapped in one rigid mode of reasoning?
5. Can you talk to a trusted colleague in a different area to help you make sense of what's going on?

On Digital

1. Can digital tools, like a meditation app or an audiobook, help you disengage and take a time-out?
2. Are you becoming overly distracted by digital tools, like your computer or phone?
3. Do you need to take a technology holiday?

FOUR

Imagination

Produce Out-of-This-World Ideas

W*HY DID THIS HAPPEN?*
Why me?

If Van Phillips were like most people, these might have been his only questions after a waterskiing accident severed his left leg below the knee. But Phillips isn't like most people. After doctors fitted him for a wood-and-rubber prosthetic leg, the young medical student and former athlete asked himself, "What if I designed an artificial leg that was as good as the original—or better? What would it look like? How would it function?"

These are the questions of an ALIEN thinker—one envisioning a new reality.

QUESTIONING THE STATUS QUO

After his accident in 1976, Phillips had good reason to simply accept the status quo and move on. For one thing, the design and functionality of prosthetic limbs hadn't changed much since World War II. Most artificial legs and feet were largely cosmetic, giving the users no energy to propel themselves. For another, his professors at the Northwestern

University Prosthetics-Orthotics Center discouraged him from even *trying* to invent a better alternative. When it came to replicating human bones, muscles, and tendons, there were technological limits to what the prosthetics industry could do. By 1984, however, Phillips would prove that the decades-long failure to develop fully functional prosthetic limbs was caused, not by technological limitations, but by a failure of imagination.

In the eight years after his accident, Phillips's goal of designing a better artificial limb became an obsession. Fortunately, he possessed several creative advantages that most industry experts did not. One was his outsider status. Even as he acquired more and more expertise in the field of prosthetics, he didn't allow his perspective to be "contaminated" by the conventional wisdom. At one point, a mentor advised him to visit the patent office to research everything that had been done on prosthetic foot inventions. His reaction? "I'm not going to pollute my mind with everyone else's ideas. I'm following my own path, not somebody else's."[1]

Moreover, Phillips was in no hurry to find solutions. Instead of searching for quick answers, he asked lots of questions and allowed these queries to send him in new and sometimes unexpected directions. At one point, while thinking about the spring force of a diving board, he asked himself, "What if you could replicate a diving board's propulsive effect in a prosthetic foot?" Later, after studying animal leg movements—particularly how the cheetah's hind legs generate powerful spring force when they are bent, he asked, "What if a human leg could be more like a cheetah's?"[2]

Eventually, this process of continuous questioning helped him to connect seemingly disparate ideas into a coherent innovation strategy. Chief among these ideas was one from a childhood memory—the image of an antique Chinese sword with a C-shaped blade, which his father had owned. When Phillips recalled that the curved blade was stronger and more flexible than a straight blade, this might have caused yet another question to pop into his mind: *What if, instead of a conventional L-shaped lower leg and foot, I created a limb that consisted of a long, continuous curve from leg to toe—like the Chinese sword?*

Given the right design and materials, I could build an artificial leg that would combine the elasticity of a cheetah's tendons with the bounce of a diving board. This would give amputees like me the ability to not only walk but also run and jump.

THE PRODUCT OF IMAGINATION

The product birthed by Phillips's imagination was the Flex-Foot.

Built from carbon graphite, which is stronger than steel and lighter than aluminum, the Flex-Foot featured a variety of designs for a range of amputees, the most famous of which was the Cheetah. After this J-shaped design, tailored specially for elite athletes, enabled Phillips to run down a hallway, he quit his job at the Center for Biomedical Design at the University of Utah to found a new company. Upon locating business partners and turning his basement into a laboratory, he experimented with one design after another, making and breaking dozens of different prototypes and then refining the models after each failure.

In 1984, Flex-Foot Inc. began marketing the new designs. By the late 1980s, however, the J-shaped model had been replaced by a C-shaped one that had no heel. By this time, Phillips had realized that the biggest mistake made by the prosthetics industry all those years had been focusing on how to replicate the human leg and foot. By contrast, his great insight had been recognizing this as a dead end. Instead of trying to mimic human anatomy, he focused on creating models—however off-the-wall—that would provide a "power source" for the amputee, using materials and designs that were *analogous* to ligaments, tendons, and bones.

Phillips's products look nothing like the limbs they're intended to replace, but they work! They work so well that his creations have been used to scale Mount Everest and enabled a double-amputee sprinter to compete in an NCAA track-and-field competition. Amputees have also used Cheetahs to run the Boston Marathon and complete an Ironman triathlon. Most famously, the South African runner Oscar Pistorius used two Cheetahs to compete in the 2012 Olympics.

Just as important for Phillips, his brainchild enabled him to run along the beach every day near his Southern California home.

In 2000, Phillips sold his company to Ossur, an Iceland-based prosthetics and orthotics company, which continues to sell the Cheetah and his other designs. Its chief executive, Jon Sigurdsson, called Phillips "a visionary, whose ideas and progressive techniques are central to our heritage."[3] Paddy Rossbach, president and chief executive of the Amputee Coalition of America, said, "Van Phillips's foot changed the whole field of prosthetics."[4]

WHAT IS IMAGINATION?

To create breakthrough solutions like Van Phillips, you need imagination.

But what is imagination, and how can you cultivate it? Most important, how can you summon this genie whenever you need it most?

As the root word—*image*—indicates, imagination is about seeing. It's the ability of your mind to produce original ideas by envisioning something that doesn't exist. In Latin, the word *imaginari* means "to form a mental picture" or "picture to oneself."[5] So when you imagine, you are forming mental pictures of a potential reality.

Compared with most people, ALIEN thinkers are better equipped to imagine new solutions. They are not constrained by the kind of intellectual baggage that most of us tote around—preconceptions, assumptions, and biases that prevent us from "seeing."

FUNCTIONAL FIXEDNESS: A BARRIER TO IMAGINATION

Of all the mental carry-on items in this baggage set, one of the most common, and pernicious, is known as "functional fixedness." This is a cognitive bias that often limits your ability to think creatively or imagine alternative uses for familiar objects or concepts.

Coined by the psychologist Karl Duncker, the term *functional fixedness* refers to the inability to realize that something known to have a particular use can also serve other ends. When you're faced with a new problem, functional fixedness blocks your ability to repurpose old tools for new uses. In his famous candle problem, for example, Duncker presented his test subjects with a box of candles, a box of thumbtacks, and a book of matches. The goal? Affix the candles to a wall using only these "tools."

For many people, the challenge involved just three items—candles, thumbtacks, and matches. An ALIEN thinker would quickly understand, however, that she had received not three tools but four: the candle box could also serve as a shelf for holding the candles. The average person's functional fixedness would likely blind them to this possibility. They would see the box as nothing but a container rather than a multipurpose tool.[6]

Another exercise, one we sometimes use as a warm-up activity in class, involves a simple jigsaw puzzle with twenty-three pieces that have a shiny red color on one side and a matte gray color on the other. But there's a twist. The puzzle can only be solved if you relax your learned assumptions about what a traditional jigsaw puzzle looks like—for example, that it has four corners, that you need to use all of the pieces, and that you must assemble them so that the shiny red color shows. When complete, the puzzle combines shiny red pieces and matte gray pieces, and none of the corner pieces are in the corners. The exercise reveals the degree to which most people allow previous experience to guide their thinking when they encounter novel situations and how this can prevent them from exploring new and unusual practices that don't seem to fit the established way of doing things.

The influence of functional fixedness tends to increase with age. The more you've practiced a solution, the harder it is to see alternatives. Young children are far less prone to this cognitive bias. In the candle problem above, researchers discovered that adults and older children (six- and seven-year-olds) are significantly slower to use the extraneous material than five-year-olds. As you gain more experience using objects, you lose this functional fluidity and become fixated on the "proper" use of different objects and concepts.

One counterexample and source of inspiration are IKEA hackers, who find new ways to repurpose IKEA furniture. A website devoted to these hacks (www.ikeahackers.net) presents an interesting array of the creative possibilities that can be produced once functional fixedness is overcome. For example, cheap, round wooden stools can be turned into a laptop table, a coat rack, or even a balance bike (no pedals or chain) for toddlers.

ALIEN thinking avoids cognitive biases. It prompts imaginative leaps by rejecting default responses and exploring novel alternatives.

In the case of the Flex-Foot, Van Phillips resisted the prosthetics industry's fixation on artificial limbs that resembled actual legs and feet. Instead, he connected the dots between objects that could provide the same (or better) functionality as human limbs—diving boards, cheetah legs, and C-shaped swords. These diverse objects had little in common with human bones and tissue, but their analogous properties helped to fuel Phillips's imagination. By recognizing alternative uses to which the designs of the sword, cheetah leg, and diving board could be put, he pursued an innovation strategy that led to a major breakthrough.

A GIFT OR A SKILL?

Shrouded in mystique, imagination is often portrayed as a gift. Like innate talent, it's frequently viewed as a quality that you either have or don't have—something you're born with (or not). While most of us are condemned at birth to be uninspired dullards, this thinking goes, a select few are granted creative clairvoyance.

This view may be justified in the art world, where aesthetics and emotions play such a big role, but not in other domains. We see imagination as a universal quality—a trait that everyone acquires at a very young age. Unfortunately, it's also something most of us progressively lose over time.[7]

Imagination is socialized out of us by our education systems. Schools, for example, are often criticized for not playing a role as catalysts of imagination. In one of the most-viewed TED Talks of all times (fifty-six million views to date), education expert Ken Robinson bemoaned the hierarchy of subjects that puts math and science at the top and art and music at the bottom, and contended that creativity should be "as important as literacy." He also noted that school "teaches" you to give expected answers, to stigmatize failure, and to fear the judgment of others. By adulthood, you have learned to suppress many of your creative capacities.[8] Is it any wonder, then, that most people try to minimize (or at least hide) their daydreaming? Who wants to be typecast as a frivolous and unproductive time waster?

To overcome the conditioning that values memorization over discovery, you need techniques to promote and revitalize your imagination—to break through the bonds of conventional thinking.

One such technique was pioneered by psychologist Edward de Bono, whose big contribution to our understanding of imagination (which he called "lateral thinking") was that it's a capability that can be stimulated and nurtured. One of his techniques involves the adoption of different "thinking hats" to force you outside your habitual thinking style and consider a problem from a more critical, positive, or emotional perspective; another technique tries to find an association between a random word (noun) and whatever challenge you face to attack it from a new angle and stimulate creative ideas.[9]

In organizations, people sometimes equate imagination with a flash of divine inspiration that emerges through some sort of brainstorming exercise. If conducted properly, brainstorming can help innovators tap into the collective creativity that exists within a group of people. In one of their courses, "Brainstorming: Rules and Techniques for Idea Generation," IDEO U (an online school created by the design firm IDEO) offers guidelines to turn brainstorming into a fruitful exercise. They include the following:

- Defer judgment.
- Encourage wild ideas.
- Build on the ideas of others.
- Stay focused on the topic.
- Allow only one conversation at a time.
- Be visual.
- Go for quantity.[10]

If these rules resemble those you might find at an improv class, that's no coincidence. Both brainstorming and improv require collective invention, and neither (in theory) has a specific end goal in mind. Instead, the objective is to produce as many original ideas as possible, without judgment or self-editing, and see where those ideas lead the group.

There's only one problem. Although brainstorming techniques promote the generation of ideas, they don't necessarily produce *original* ideas. And when they do, the most radical ideas are often deleted or diluted fairly quickly. The harsh reality is that creative-thinking efforts often generate the illusion of imagination but yield disappointing results.[11] For this reason, Jake Knapp, a designer who cofounded Google Meet and helped build products like Gmail and Microsoft Encarta, banned brainstorming exercises from his celebrated design sprint process.[12] Instead, he favored having each person sketch a few critical ideas on paper with words and pictures to develop a coherent concept and encouraged in-depth critical thinking by each individual on the team. No more shouting over one another's ideas. Each person must work independently and take the time to think through their ideas and express them clearly so they can be pitched to the rest of the group.

Methodology alone is not enough. Creativity also depends on mindset. To achieve the promised benefits of brainstorming—whether collectively or individually—you need to both *release* your imagination and *stimulate* it.

RELEASE YOUR IMAGINATION

Because imagination is often held prisoner by perceived constraints, whether social (fear of failure or ridicule) or cognitive (functional fixedness regarding objects, processes, or concepts), to fully exercise your imagination, you must first unlock the cell door. To do this, ALIEN thinkers approach situations and problems as children do—with an open mind.

This is how an Argentine auto mechanic invented a new medical device that could save hundreds of thousands of lives each year.

For many years, Jorge Odón had tinkered with new inventions in his garage, but unsurprisingly, these creations were all car related. One day in 2006, however, he noticed several employees reenacting a YouTube video in which a cork that had been shoved inside a bottle was retrieved with a plastic grocery bag. The bag was inserted into the bottle, inflated until it surrounded the cork, and then drawn out. Later that night at dinner with friends, Odón defied his friends to recover a cork from inside an empty wine bottle and won a bet by demonstrating the feat.

But that wasn't the end of it. According to Odón, he woke up his wife at 4:00 a.m. the next morning to tell her that he'd just had an idea: What if he invented a device, similar to the cork remover, that could free a baby stuck in its mother's birth canal?

His wife said he was crazy and told him to go back to sleep.

Although his own children were born without problems, one of Odón's aunts had suffered nerve damage during her birth. And she was far from alone. Every year, about 10 percent of the 137 million births worldwide have potentially serious complications. Nearly six million babies are stillborn or die shortly after birth, and more than 250,000 women die in childbirth every year; maternal mortality rates are especially high in poor nations. Obstructed labor, which can occur when an infant's head is too large for the birth canal, is a major factor contributing to the mortality rate.

Until recently, the most common tools used to extract babies were forceps (essentially, large pliers) and suction cups, which are attached to the baby's scalp. Both tools carry the risk of crushing the baby's head, twisting its spine, or causing hemorrhages in either the infant or the mother.[13] By contrast, with the Odón device, these risks vanish. An attendant slips an inflatable air cuff inside a lubricated plastic sleeve around the baby's head, inflates it to grip the head, and then pulls the sleeve until the baby emerges.

Later the same morning that he had the idea for the device, Odón was introduced to an obstetrician. The doctor was more encouraging than his wife, so Odón kept working on his idea. He built the first prototype in his kitchen, using a glass jar for a womb, his daughter's doll for a trapped baby, and a fabric bag and sleeve as the lifesaving device.

With the help of a cousin, Odón then met the chief of obstetrics at a major hospital in Buenos Aires. The chief had a friend at the WHO who knew Dr. Mario Merialdi, who was in charge of the WHO's efforts to improve maternal and perinatal health. At a 2008 medical conference in Argentina, Dr. Merialdi gave Odón ten minutes to present his idea. Ten minutes turned into two hours. By the end of their meeting, Dr. Merialdi was impressed. He arranged for testing at a simulation lab. Odón has been refining the device ever since.

"This problem needed someone like Jorge," said Dr. Merialdi. "An obstetrician would have tried to improve the forceps or the vacuum extractor, but obstructed labor needed a mechanic. And ten years ago, this would not have been possible. Without YouTube, he never would have seen the video."[14]

The Odón device has the potential to save hundreds of thousands of women and babies and reduce the number of caesarean-section births in wealthier nations. The US-based medical technology company Becton Dickinson acquired the rights to manufacture the device and started clinical studies in 2016. The results triggered modifications to various features (including the applicator and handles), and in 2018, the improved device entered phase two of testing. All because a car mechanic released the hand brake on his imagination.

Seek a Playful State of Mind

Another way to generate new ideas is to adopt a playful state of mind. It was Odón's playfulness that prompted him to try the cork trick on his friends and then challenge them to work out the solution for themselves.

Play has long been recognized as integral to imagination. Even before researchers began cataloging its benefits, some of history's greatest imaginations demonstrated its utility. As biologist Patrick Bateson noted in his article "Playfulness and Creativity," many of the most creative composers, artists, and scientists were quite playful. Mozart was infamous for his impish sense of humor, which is reflected in some of his music. For example, his three-voice canon (KV559) consists of nonsensical Latin lyrics that, when sung, sound like risqué German phrases. M. C. Escher said of his designs, "I can't keep from fooling around with our irrefutable certainties. It is . . . a pleasure knowingly to mix up two- and three-dimensionalities, flat and spatial, and to make fun of gravity." Even the discoverer of penicillin, Alexander Fleming, was accused by his boss of approaching research like a game. When asked what he did, he replied: "I play with microbes. . . . It is very pleasant to break the rules and to be able to find something that nobody had thought of."

Because play is about breaking away from established patterns and combining actions or thoughts in new ways, it helps promote creativity—creativity also involves breaking free of conventional ideas and behavior. This is why creative people are able to detect previously unseen patterns and relationships, and to connect seemingly disparate elements into new forms.[15]

Steven Johnson, the author of *Wonderland: How Play Made the Modern World*, would agree. He argues that necessity is not the only mother of invention. Play is also an important parent, and we have seriously underestimated how play has shaped the world.

"When we look at models of what drives historical change and innovation in society," he said,

> we think about necessity and the quest for power as the traditional prime movers of change, but I argue that the world of play, delight and wonder—things we did just for the fun of it—actually end up transforming society in ways that are really profound and that set the groundwork for really transformative ideas.
>
> Going back in early technological history, some of the earliest things that people built were bone flutes. We think they are at least 70,000 years old. So imagine back then, you had a whole world of

things to invent. You already have spears and sewing needles and a few things like that. You can invent anything you like—and what do you choose to invent? You invent a flute. There's no function to it, but there's something about the sound that is pleasing and interesting to our ears.[16]

In sum, we humans have always sought out intriguing and pleasurable new experiences, often with the assistance of technology.

Because play is exploratory and nonjudgmental, it can prompt you to take more chances—to boldly embark along paths that your prudent self would normally avoid. Fortunately, your playful self doesn't worry about following a path that may lead to a dead end. It's focused on the discovery process, not so much on whether it reaches a particular destination. Your playful side has fun blazing new trails, whether they lead to breakthrough solutions or a brick wall.

For these reasons, playful people—daydreamers, stargazers, and those with their heads in the clouds—are more likely to devise original solutions. For them, work is play. This attitude gives them the emotional space they need to make judgment-free mistakes and then study the results of those mistakes with an open mind.[17]

When preparing for fast-iteration innovation exercises (deep dives), we often ask participants to draw their neighbors in just two minutes. The aim is to remind them that they have to be playful to create: to set aside their concerns about the judgment of others or about appearing ridiculous, as well as the feeling that they don't have the competence needed or that the deep dive is a trivial exercise.

Young children don't have these inhibitions. They realize that their drawings will not be perfect representations of reality. They're carefree and throw themselves into it, unconcerned about whether the picture is pretty or whether others may laugh.

The drawing exercise is a metaphor for the type of mindset that you need in order to innovate. As George Bernard Shaw once said, "We don't stop playing because we grow old; we grow old because we stop playing."

Brainstorm Questions, Not Answers

If a playful attitude supplies the oxygen needed to ignite the fires of your imagination, brainstorming for questions provides the kindling.

But the process can be trickier than it sounds. Most of us have been conditioned to seek answers, not to pose questions for which we have no ready answers. Because we've been trained to find solutions, lingering in a state of uncertainty can make us very uncomfortable.[18] Yet uncovering the right questions, especially provocative questions, is necessary for stimulating individual and collective imaginations—for opening up new perspectives and unlocking the doors to less-than-obvious solutions.

In general, brainstorming frameworks focus on generating a large quantity of questions—however outrageous, silly, or seemingly unrelated to the challenge. Some impose a strict time limit on the question incubation period as a way to force participants to keep the questions coming instead of pausing to self-edit or critique other people's questions.

In addition, it's widely recognized that certain types of questions tend to produce better results than others. For example, design thinking insists on the importance of asking, "How might we . . . ?" (HMW) questions as a way to focus brainstorming exercises. HMW questions are useful because they help frame or reframe the problem you're trying to solve. However, they cannot guarantee imaginative solutions because they do little to remove the mental constraints (such as functional fixedness) mentioned earlier.

ALIEN thinkers understand the importance of brainstorming game-changing questions that unlock the imagination and open up the search for new possibilities. They dare to ask disruptive, transformative, and even uncomfortable questions for which people have no easy answers. In some organizations, these provocative questions might be considered heretical because they seek to remove tacit constraints on what can be said or even thought. They may raise eyebrows because they reflect the same kind of naivete that children demonstrate in their interactions with the world, which most adults and experts are desperate to avoid.

Several types of questions help the "provocative inquiry" that promotes breakthrough thinking, but the top two creativity catalysts are these:

Why Questions

These are the first questions that children ask to understand how the world works. "Why do I have to go to bed now?," "Why can't I fly?" When it comes to uncovering all the innovative possibilities that may exist, it is critical to ask why questions, especially if you wish to challenge the status quo.

This points to another limitation of traditional brainstorming exercises: they get people to shout ideas that sound right regardless of how well the initial inquiry was framed. We generate as many answers to the question that was posed as possible, without giving much thought to how relevant that question was in the first place. And that is a mistake. Sometimes it's simply the wrong question to ask, or there may be more thoughtful ways to frame the problem. The more questions we ask, the more curious we become and the more engaged we are in the pursuit of innovative solutions to the problems we face.[19] Examples of why questions that you could ask, at work or at home, include these:

- Why does a particular situation exist?
- Why does it present a problem or create a need or opportunity, and for whom?
- Why has no one addressed this need or solved this problem before?
- Why do you personally (or your team or organization) want to invest more time thinking about, and formulating questions around, this problem?

This is the approach Van Phillips took at the beginning of his journey to design a better prosthetic limb. Instead of asking "Why me?" he wondered why artificial legs should be made of wood and rubber, why a prosthetic foot should be made to resemble a human foot, and why the focus shouldn't be on performance instead of appearance.

What-If Questions

What-if questions help you explore the possibilities and dare you to uncover something different or unusual, without care for feasibility or possible ridicule. For example, Jorge Odón asked himself, "What if the same principle that can retrieve a cork from inside a bottle could save a baby stuck in the birth canal?"

The equivalent question for an organization could be, "What if we no longer did what we do now?" In 2009, the McLaren Group, best known for Formula One racing, asked itself that very question. Formula One cars are filled with sensors, and it was McLaren's job to leverage the data to create models and strategies that would lead the drivers to victory. As soon as McLaren employees asked what-if questions about *other* potential uses for such data, they were able to leverage McLaren's performance-improvement capabilities for a wider range of clients—from elite sports teams to health-care systems and air traffic control services. Since 2009, McLaren Applied has designed health-monitoring systems for stroke victims and amyotrophic lateral sclerosis patients based on Formula One telemetry. It has also created a scheduling system for Heathrow Airport that reduces flight delays, and has worked with some of the world's biggest consumer goods companies.

What-if thinking liberated the top team to think about how McLaren's world-class capabilities in material science, aerodynamics, simulation, predictive analytics, and teamwork might apply to other sectors. This division has become the fastest-growing and most profitable part of the group, paving the way for McLaren to morph into a consulting and technology group that happens to have a successful Formula One team.[20]

STIMULATE YOUR IMAGINATION

Besides shedding your social and cognitive inhibitions, you can also stimulate your imagination through new inputs and associations.

Take the case of Chris Sheldrick, founder of What3words. A combination of his deep-felt frustration with GPS and his experience as a chess player inspired a more reliable alternative. Together with a mathematician friend (a former teammate on his school's chess team), he divided the world into a grid of three-meter squares, each with its own unique three-word identifier. This breakthrough solution has proven especially valuable for delivering mail or medicines to people without addresses—for example, in remote villages and shantytowns.

The idea for What3words came to Sheldrick when he was running a live-music booking and production agency, where he remains a non-executive director and which shares an office with his new company. Logistics were a big part of his job, and Sheldrick found that if he was sending directions and addresses to thirty people, there would always be a phone call or two from people saying things like, "I'm under a lamp by a hedge. Is that right?"

"Postcodes are fine for some places," said Sheldrick, "but they are not much help if you are going to somewhere like Wembley Stadium, which has one address but twelve car parks.

> Over a cup of tea with a mathematician friend, I shared my frustration with GPS coordinates and strived to find a way of naming everywhere in the world that would be incredibly easy for people. We worked out that a list of 40,000 words was enough to give every 3m x 3m on the planet its own unique 3 word address. We subsequently wrote an early version of the what3words algorithm on the back on an envelope.
>
> There are a number of systems out there, but everyone else has solved the problem in exactly the same way. People have reduced coordinates into alphanumeric codes, but these comprise far too many characters to be realistically usable. Replacing 16 numbers with a combination of 9 numbers and letters just wasn't a proper solution in our eyes, and [they] are near impossible to remember. A person's ability to retain 3 words in short term memory is almost perfect. Anyone can get their head around 3 words.[21]

Today, What3words is used in more than 170 countries by individuals, businesses, aid agencies, and emergency services. The company also has partnership agreements with five national postal services and has been used to coordinate security at the Glastonbury Festival, Burning Man, the Super Bowl, the Olympics, and the World Humanitarian Summit. In addition, the system is being built into self-driving vehicles and drone delivery systems, and ordinary people are using it to meet friends or remember where they parked their car.

This example illustrates two established ways of igniting your imagination: using analogies and combining concepts.

Using Analogies

So-called analogical thinking encourages your brain to make different types of connections. It prompts you to draw unexpected parallels between two distinct things. Sheldrick and his school friend had been members of the Eton chess team together. They quickly gravitated toward a chessboard configuration with easily identifiable squares as inspiration for the imaginary grid that now covers the entire planet—land, sea, and ice caps included.

You can use analogies to generate new ideas or consider a problem from a fresh angle. They encourage your brain to transfer information from a domain you understand to help resolve a challenge in an unfamiliar area.[22] Psychologist Dedre Gentner calls it "bootstrapping the mind."[23] You can find helpful analogies not only in human constructs but also in nature. For example, the reflective property of cats' eyes inspired Englishman Percy Shaw to develop reflectors to help motorists drive safely at night, and the annoying thistle burrs that attach themselves to clothing on hikes spurred Swiss engineer George de Mestral to invent Velcro.

When you are trying to come up with new solutions to a problem, analogies can open up new thinking spaces. Ask yourself, "What else is like this problem?" or "Where have I seen something like this before?"

Analogies disregard the conventional boundaries between knowledge domains. Consider the example of N. Joseph Woodland. He took on the challenge, set by the owner of a local supermarket chain, of automating the transaction-recording process. Two analogies helped him crack the problem. To represent the information visually, he felt he would need something like Morse code (which Woodland had learned as a boy), with its simplicity and limitless permutations. The other inspiration was a pattern he absentmindedly drew in the sand while relaxing on the beach. Woodland later recalled, "I poked my four fingers into the sand and for whatever reason—I didn't know—I pulled my hand toward me and drew four lines. I said, 'Golly! Now I have four lines, and they could be wide lines and narrow lines instead of dots and dashes.'"[24] The furrows of different widths were like a graphical equivalent of Morse code. And that is how he came up with the concept of the bar code. Although he patented the invention in 1952, it proved too costly to implement with existing technology. It took several decades of advances in scanner and computing technology to unleash the full power of the bar code.

In groups, analogies trigger more creative conversations. They are powerful tools for achieving a common understanding of complex problems and generating new insights. Researchers from McGill University in Montreal studied scientists working in four microbiology labs and found that they used up to fifteen analogies in a one-hour lab meeting. And the more successful labs employed more analogies when discussing their work.[25]

Combining Concepts: Flying Donkeys

While analogies inspire creativity through figurative associations, creativity also relies on more literal combinations of ideas or disciplines. In the case of What3words, the association of mathematics and language was clear at the outset. The cofounders quickly worked out that a list of forty thousand words would give around sixty trillion combinations, enough to give every three-meter-by-three-meter square on the planet its own unique three-word address.

Creativity is often the product of odd combinations. At the most basic level, you can import ideas from other worlds and apply them to your own. "Stealing with pride" means transforming the original concept and giving it new life. Through attention or levitation, you may have noticed interesting insights and practices from one domain that can be adapted to another.

Jonathan Ledgard, a war reporter and longtime Africa correspondent for the *Economist*, imagined a drone-based network that would deliver blood and medical supplies to remote areas across Africa.

During his travels in Africa, Ledgard witnessed many social and economic challenges (unconnected dots) that other Western journalists had already documented, including these:

- Economies that are largely preindustrial but whose populations are equipped with digital communications technology (internet-enabled mobile phones, tablets, etc.)
- High youth unemployment. According to a World Bank estimate, 80 percent of African young people will be unable to find salaried work over the next decade.
- Poor transportation infrastructure. In parts of Africa, moving physical objects is as daunting today as it was for medieval Europeans.

A key difference between Ledgard and other visitors to Africa was that he made unique connections between these dots and others he uncovered through conversations with people in high-tech fields, especially robotics. Among the facts he connected to the above challenges were (1) that drones can transport more than just missiles and surveillance equipment, and (2) that the cost of the robotics in drones is plummeting to the point that they will soon be affordable for even the poorest nations.

Armed with this knowledge, Ledgard envisioned a drone-based air cargo venture that would deliver modest payloads between Africa's smaller cities and towns—drones flying back and forth between the same destinations dozens of times a day with cargos of twenty to forty kilograms. Shortly after developing this vision, Ledgard described the idea to a livestock farmer he met in northern Kenya.

> We really want to create this vehicle and we want to put maybe twenty kilos of payload into it, and you can fly it into the sky and then it can drop off something and pick up something. And he was having real difficulty understanding what I was talking about. But then he leaned back, nodded, and said, 'Oh, I know. You want to put my donkey in the sky.'
>
> And we realized, yes, in a way that's exactly what we wanted. We wanted a drone that was going to be a midsized vehicle, donkey-size, which would carry pretty much the same payload as a donkey and travel a little bit farther—a 150-kilometer range. So we decided the drop the word *drone* and use *donkey*. We are developing flying donkeys.[26]

Thus was born RedLine, which began trial operations in Rwanda in 2016.

THE OUTSIDER ADVANTAGE

Several of the breakthroughs we've discussed in this chapter illustrate this type of combinatorial thinking. In addition, these cases involve people who were outsiders in the field to which they contributed. Sheldrick was not a localization specialist. Odón was not an obstetrician. Van Phillips was not an engineer.

This is no accident. Research shows that outsiders often find it easier to reason like ALIEN thinkers and develop novel solutions. In many cases, outsiders are able to better connect disparate thoughts because they come to the table with fewer preconceptions than insiders.

In one research study, for example, separate groups of carpenters, roofers, and in-line skaters were asked for ideas on how to improve the design of carpenters' respirator masks, roofers' safety belts, and skaters' kneepads. Independent assessment of the solutions showed that each group was significantly better at coming up with novel solutions for the fields *outside* their own.[27]

Moreover, while examining 166 problem-solving contests posted on the InnoCentive innovation platform, Harvard Business School professor Karim Lakhani found that the winning entries were more likely to come from "unexpected contributors" with "distant fields" of expertise who are foreign to the focal field of inquiry.[28]

Confirming this "advantage of marginality," a separate crowd-sourcing study revealed that industry outsiders were more likely than insiders to come up with breakthroughs solutions to relatively complex and intractable R&D problems. (But outsiders also needed to invest a disproportionate amount of time and effort to achieve this "big C" creativity.)[29]

THE MEDICI EFFECT

Organizations can spark similar connections by assembling minds with diverse knowledge bases and perspectives to achieve the "Medici effect," a term coined by author Frans Johansson.

The Medicis were an influential family of bankers and politicians in Renaissance Italy. They are best known today as patrons of such diverse artists, architects, and philosophers as Michelangelo, Leonardo da Vinci, and Niccolò Machiavelli. By bringing together the greatest talents from across Europe and as far away as China, said Johansson, the Medicis were able to break down the boundaries between a variety of disciplines and cultures, touching off an explosion of creativity.

Essentially, the Medici effect refers to how diversity drives innovation. By assembling a heterogeneous group of people with different viewpoints, backgrounds, and talents, you enable everyone to borrow and build on everyone else's ideas to break new ground. People and organizations wishing to jump-start their innovation efforts should strive to create the same diversity among their people as among those the Medici family sponsored.[30]

DIGITAL IMAGINATION AND CREATIVITY

Can machines help humans be more imaginative?

More intriguing (or alarming), can machines *themselves* be imaginative and creative?

When it comes to assisting human imagination, digital tools can help stimulate lateral thinking and imaginative connections. For example, one search engine, Seenapse, brings together disparate parts of the internet to stimulate creativity. It does this by capturing and sharing other people's logic leaps among search results to stimulate your own. Where they have been may inspire you to visit locations you would not normally have considered. Another search engine, Yossarian, generates metaphors of varying degrees of separation from the search term to help people make creative leaps and associations.

Digital tools also help you join the dots between remote studies in separate fields. For example, a company called BenevolentAI is using artificial intelligence to mine and analyze enormous quantities of biomedical information—from clinical trial data to academic papers. Among other things, BenevolentAI can identify molecules that were used to treat certain illnesses during clinical trials, whether successful or not, and then predict how the same compounds can be used to treat other diseases.[31]

In January 2020, scientists at the company turned their algorithm to the new coronavirus emerging from China. The company trawled its large respository of results from medical journals in search of existing medicines that could be used to treat the virus.

Joanna Shields, BenevolentAI's chief executive, told reporters, "Rather than focusing solely on drugs that could affect the virus directly, we explored ways to inhibit the cellular processes that the virus uses to infect human cells."[32] Within ninety minutes of analysis, a potential treatment emerged: an oral drug primarily used to treat rheumatoid arthritis, marketed by US pharamceutical company Eli Lilly. The algorithm predicted that the drug could block the virus's ability to replicate and enter human cells. It also indicated that the medication could reduce the severity of symptoms experienced by those infected with the virus.

Scientists from BenevolentAI contacted Eli Lilly, and the drug-maker quickly conducted tests of its own, which largely confirmed BenovalentAI's findings. Eli Lilly then launched a set of clinical trials to further test and validate the drug's effectiveness in treating COVID-19.

BenevolentAI can also use the predictive power of its AI algo-rithm to design new molecules, extracting a new hypothesis based on a knowledge graph composed of over a billion relationships between genes, diseases, proteins, and drugs. "When the periodic table of the elements was generated, there were gaps in that table where you know elements had to exist, but they hadn't been discovered," said Jackie Hunter, CEO of BenevolentBio, a division of BenevolentAI. "We use our knowledge graph like that: what relationships should be present but are not yet known?"[33]

Machines can also assist the human imagination by *asking* ques-tions as well as answering them. This can be accomplished through a process called "automated hypothesis generation," in which advanced computers sift through massive quantities of information looking for nonintuitive links. From these links they propose hypotheses that can be further refined and tested by humans. Today, nearly one hundred groups of scientists are working to develop tools to automate the pro-cess of hypothesis generation. Their intention is to use these tools to trawl through the hundreds of millions of periodicals, theses, and academic papers that sit in databases around the world. This suggests that automated hypothesis generation, not artificial intelligence, may be the technology that inspires future breakthrough innovations.[34]

Digital anonymity is another feature that can facilitate creativity and imagination. Many digital tools and applications allow for anon-ymous browsing and participation, which can free people from worry about others' perceptions, allowing them to pursue imagination and creativity with less reticence.

Digital tools and technologies can help humans release and stim-ulate their minds by giving them new tools to play with, by asking different questions, by proposing analogies, and by joining the dots between distant fields.

RISE OF THE MACHINES?

Can machines *themselves* be imaginative and creative?

Most conventional wisdom suggests that computers are optimized for routine, structured tasks. Although they can accomplish these tasks with greater speed and precision than humans, tasks that require creativity, innovative thinking, and imagination are largely beyond their capabilities. These "higher-order tasks" are solely within the realm of human capabilities. Even when a computer beats a human in chess, it's not because the computer is more creative. The computer wins because chess is structured, bounded, and predictable, and the computer can simply outcalculate the human.

But if the conventional wisdom is really true, then how is it possible that computers can also beat humans in Texas Hold 'Em, a form of poker in which bluffing and reading your opponents is as important as raw processing power? And what about the game Go, which is so complex that even the world's largest supercomputer couldn't get close to figuring out all the possible moves and permutations?

Ada Lovelace, one of the inventors of modern computers, was also one of the first people to see the potential for computers to do more than simply calculate numbers. The daughter of the poet Lord Byron and mathematician Anne Isabella Milbanke (a powerhouse marriage of art and science), Lovelace worked with computer pioneer Charles Babbage to develop the first computer, called an analytical engine, and understood that the algorithms used for computation could be applied to fields other than mathematics. For example, she postulated that if music or art could be broken down into rules, then symbolic logic (i.e., algorithms) could be used to program the analytical engine to create pictures or musical scores. She even wrote about building a machine to compose music.[35] But as optimistic as she was about the power of computers to create art and music, she nevertheless maintained that machines could not easily reproduce creativity and imagination.

More than one hundred years later, in the 1950s, computer pioneer Alan Turing believed that computers would soon think like humans. He created the Turing test to prove it, which has since been passed. However, some people now prefer the Lovelace test to the Turing test. The Lovelace test requires computers to prove creativity as well as analytical ability.[36] In particular, to pass the Lovelace test, an artificial agent must be able to produce and reproduce a creative output, perhaps an idea or a piece of music, that it was not engineered to produce. Further, the agent's designers must not be able to explain how their original code led to this new output. The Lovelace test has yet to be passed by any artificial system.[37]

In recent years, strong evidence has emerged that computers are going beyond what we tell them to do, exhibiting behaviors that look very much like creativity and imagination. And as computational power increases, there's every reason to believe that machines' ability to think creatively will increase.

One example is the famous "move 37" that Google DeepMind's supercomputer AlphaGo made against world Go champion Lee Sedol in their second match in 2016. Move 37 shocked the Go community because it was so counterintuitive, contradicting a thousand years of orthodoxy. It broke many of the accepted rules of Go, like playing close to the edges and dominating a particular part of the board before moving to other parts. Only later in the game, many moves later, did move 37 become significant. In fact, it helped the computer beat Lee Sedol in that game and 4–1 in the series.

There was no way the computer could have calculated that many moves into the future. The permeations were just too numerous to simulate. Discussing the pivotal move, Demis Hassabis, the founder of DeepMind and a former chess prodigy, noted, "In some sense, AlphaGo knew that this move 37 was very unusual, because it gave the probability that a human would play that move as *one in ten thousand*. So it wasn't just learning and copying what humans do; it was actually innovating."[38]

Looking back on the series, Hassabis said, "This test bodes well for AI's potential in solving other problems. AlphaGo has the ability to look globally across a board—and find solutions that humans either have been trained not to play or would not consider. This has huge potential for using AlphaGo-like technology to find solutions that humans don't necessarily see in other areas."[39]

Indeed, DeepMind has also been working on using AlphaGo's imagination and creativity to solve problems outside of gaming. For example, it figured out how to reduce power consumption in Google's data centers by 15 percent, resulting in savings of hundreds of millions of dollars.[40]

The secret behind AlphaGo and many other computers that appear to show creativity and imagination is the use of a nonstructured approach to computation called neural networks. Neural networks are not new. They've been around in one form or another since the 1960s. The idea behind them is to roughly simulate the way that biological systems work, particularly brains, where multiple streams of data are mathematically weighted and analyzed iteratively through a series of layers.

Neural networks regularly come up with outputs that are qualitatively different from inputs. These outputs frequently appear to be creative and imaginative, but while neural networks simulate the processes of human brains, they don't work in exactly the same way. However, the process is less important than the output. If both humans and machines can come up with ideas that are creative and imaginative, does it really matter how the ideas come about?

PUTTING ALIEN THINKING TO WORK
Stora Enso's Creative Block

An imagination challenge we observed at close quarters involved the Scandinavian paper giant Stora Enso briefly mentioned in Chapter 2. Once a leader in its field, by the mid-2000s, the company had hit a wall.

With the shift from print to online publishing and the shrinking demand for paper, the top team spent four difficult years, starting in 2007, engaging in several rounds of cost cutting—divesting mills and laying off workers—to stabilize the situation. It worked. But then they faced the challenge of shifting gears to innovate and create new growth markets.

In early 2011, the nine-person team—all male, all Nordic, all paper industry veterans—realized they were ill-equipped to ask the sort of heretical questions needed to reinvent the company. Jouko Karvinen, CEO at the time, recalled a particular meeting: "I sat there listening and realized that all of us were telling the same old stories, over and over again."

Karvinen understood that they needed to bring diverse perspectives into the conversation. But instead of taking the conventional route of hiring consultants with ready solutions, he approached us to help Stora ask better questions, boost its imagination capability, and cocreate a solution.

Reimagining the Future

The initial insight was to establish a "shadow cabinet," drawing on the next generation of leaders at Stora Enso to help the top team challenge their existing assumptions and envision new opportunities. After a series of discussions with IMD, however, the top team agreed to a more radical approach. Why focus on only the usual suspects for help? Why not open up the opportunity to all employees? As former HR chief Lars Häggström put it: "We wanted to have people who were passionate about pushing the boundaries and who question[ed] literally everything."[41]

With that in mind, they posted an ad on the company's intranet inviting all comers to apply to the Pathfinders program. It triggered 250 applications, and after a battery of assessments and interviews, sixteen people were finally selected. Significantly, they included recent hires as well as established employees, and they represented a much broader mix of demographics, hierarchical levels, experiences, and personalities than would normally have comprised this team. Several of those selected were not even on the radar of the company's talent-screening system. Though the group was eclectic, its members shared a genuine appetite for change and were far better suited to think laterally than the incumbent decision-makers.

To help *feed* their minds and enrich their perspectives, they were sent across the globe for six weeks, from China and India to the US and Latin America. To help *free* their minds, they were symbolically fired from the company—then rehired with a new mandate to challenge the company's mindset and ways of working. IMD also created a custom program to expose them to new conceptual frameworks and enable them to conduct deep market-insight studies when visiting companies.

The name Pathfinders, a nod to the Mars space exploration mission, gave the group a clear identity and captured its mandate, which was to discover new ways forward—to explore global trends and alternatives without being weighed down by Stora's long heritage, to bring back insights from companies inside and outside their industry, and to identify opportunities that were falling in between the company's silos.

Said Karvinen: "I want a revolution. I don't want PowerPoint presentations giving advice about what we could do. I want them to come back with ideas that we can implement . . . to start up a new business."[42]

Ultimately, the Pathfinders engaged the top team in profound and impactful conversations and made many strategic recommendations that were adopted. Their input proved so valuable that the initiative was renewed on an annual basis and redubbed "Pathbuilders."

One recommendation that was acted upon was related to where the new growth markets should be. While sustainability was already an area of focus, the Pathbuilders pushed for sustainability and renewable materials to be at the very center of the strategy, claiming that there was significant growth potential in this space. To grow this area faster, they made several recommendations, like putting a head of sustainability on the executive committee and accelerating R&D for renewables both internally and with partners. This contributed to Stora Enso's later making "the renewable materials company" its brand. In Pathbuilders programs in later years, new participants were tasked with building the company's future direction and coming up with ideas that would continue to push the boundaries of the organization, address pressing business challenges, develop new businesses, and shake up the company's internal organization.

Breakthrough Changes

Eight years later, the Pathfinder/Pathbuilder program's impact on innovation has been immense, transforming Stora Enso from a traditional paper-and-board producer to a global renewable materials company. Its strategic focus has shifted to new value-creating areas like fiber-based packaging, innovation in biomaterials, and bio-based chemicals. In the same period, its share price has tripled, and two-thirds of sales (and three-quarters of profits) now come from the growth businesses.

Beyond the focus on results, the program has also reshaped the entire company culture—as evidenced by surveys showing that employees are more engaged and innovative than ever before. Moreover, the self-selection mechanism reenergized the workforce. It was seen as a sign of trust in the employees and a chance for anyone to have an impact. Of those who served as Pathfinders or Pathbuilders, more than 70 percent were promoted or changed positions within six months of completing the program.

This is an innovation legacy that is now built into the company's future leadership.

KEY TAKEAWAYS

- Imagination is about *seeing*. It's the ability of your mind to produce original ideas by envisioning something that doesn't exist.
- Of the many barriers to original thinking, functional fixedness is one of the most common. This cognitive bias limits your ability to think creatively or imagine alternative uses for familiar objects or concepts—as in Karl Duncker's famous candle experiment.

- ALIEN thinking avoids cognitive biases and prompts imaginative leaps by rejecting default responses and exploring novel alternatives. For example, ALIEN thinker Van Phillips resisted the prosthetics industry's fixation on designing artificial limbs that resembled actual legs and feet. Instead, he focused on connecting the dots between objects that could provide the same (or better) functionality—diving boards, cheetah legs, and a C-shaped sword.

- Imagination is a universal quality—a trait everyone displays at a very young age. Unfortunately, it's also something most of us progressively lose over time. It is socialized out of us by our education systems.

- To overcome the conditioning that values memorization over discovery, you need techniques that promote and revitalize your imagination. ALIEN thinkers approach problems with a childlike open mind—as Jorge Odón did. They also recognize the value of play in stimulating the imagination, because it is exploratory and nonjudgmental.

- The most effective brainstorming seeks to raise questions, especially why questions and what-if questions.

- You can also stimulate your imagination through new inputs and associations.

- Analogical thinking encourages your brain to make different types of connections. It prompts you to draw unexpected parallels between two distinct things, as What3words founder Chris Sheldrick did when he drew a chessboard-like grid over the surface of the earth.

- Outsiders often find it easier to think like aliens and develop novel solutions. They are better able to connect disparate thoughts because they come to the table with fewer preconceptions than insiders. Sheldrick was not a localization specialist. Odón was not an obstetrician. Phillips was not an engineer.

- Organizations can spark imaginative connections by assembling minds with diverse knowledge bases and perspectives to replicate the Medici effect.

QUESTIONS TO ASK YOURSELF

1. Are you comfortable asking questions to which there are no immediate answers?
2. Can you use an analogy to describe a situation or problem you're currently trying to solve?
3. Have you spent enough time searching for interesting developments in other fields or disciplines that could shed new light on your project?
4. If you were trying to explain your innovation quest to a child, what would you compare it to?
5. Have you ever come up with innovative solutions in a field where you were an outsider? If so, what was the idea that helped you develop the solution?

On Digital

1. Can you use digital tools to promote creative thinking?
2. Can you use digital tools, like AI, to find hidden patterns in the data?
3. Can you read, watch, or listen to content—like a TED Talk, podcast, or audiobook—on a topic that is far away from your normal areas of focus?

FIVE

Experimentation

Test Smarter to Learn Faster

G ETTING FIRED MAY NOT HAVE BEEN THE BEST THING THAT ever happened to Laurence Kemball-Cook, but it was a fortuitous moment for the clean energy industry.

While he was a student at Loughborough University in England, Kemball-Cook spent a one-year internship at a large European energy company, where he was asked to design a solar-powered streetlamp. "I spent a year building it, but I failed," he said.

"In cities, there are lots of skyscrapers and lots of shading, so you need lots of big panels, which [aren't] very easy to fit on a pole. After a year, I got fired, but I kept thinking about the problem of energy in big cities, and I kept trying to think of another power source." One morning, while walking through Victoria railway station in London, "I thought about how busy the subway is, and I thought, 'What if we could harness the energy of human footsteps?'"[1]

BREAKTHROUGH FROM BREAKING AND ENTERING

After returning to the university in 2008 to complete his final year, Kemball-Cook began sketching a floor tile that could generate electricity from the kinetic energy produced by people's footsteps. After converting his dorm room into a workshop, he built his first 3D model out of duct tape and wood in about sixteen hours.

Designing the tile was relatively fast and painless, but convincing investors to believe in the technology was not. "It sat in my bedroom for 3 years. The government said it wouldn't work. About 150 VCs all said to me, 'It will never work.' My university said, 'Give us 75% of the company, and we'll help you.' No way. So I was stuck. I was going to give up."[2]

But he didn't.

Instead, he conducted a radical experiment. Late one night, he carried a handful of the prototypes to a building site, broke into the facility, cemented the tiles in place, connected them to the lighting, and took photos of his accomplishment/crime. The next morning, he posted the photos on the homepage of his website with the headline, "The Future of Energy Is Here!"

It was not just any building site. His target was the largest urban mall in Europe, located next to Olympic Park in London, site of the 2012 Olympics, and it was being built by the Westfield Group at a cost of £1.6 billion. The building's owners were not pleased, but the trial produced the desired results for Kemball-Cook and his company, Pavegen. "[Investors] saw that it was real, and I proved that it worked."[3]

He started receiving emails from people who wanted to buy or distribute the kinetic tiles, or invest in Pavegen. In fact, the Westfield Group itself came to recognize the value of the tiles and signed a £200,000 agreement to ensure that its new shopping center would become the first large-scale installation once the tiles became commercially available.

Almost overnight, Pavegen was transformed from a cash-starved nonentity with no proven product into an exciting investment opportunity—with enough funds to finance Kemball-Cook's ongoing experimentation.

Soon he tested a second installation. This time it was at his old school, and although he did have permission from the school, he didn't have authorization from safety and standards inspectors. The retiled corridor provoked a stampede to use it. This was a product that people really wanted to interact with.

From there, the start-up was able to raise £350,000 through London Business Angels in 2012, and it started gaining momentum. Small, temporary installations at large events, such as a four-day music festival on the Isle of Wight and the meeting space at the World Economic Forum in Davos, secured growing media attention. By 2015, awareness of the company was such that it attracted over one thousand investors in one week, raising a total of £2 million on the British crowdfunding platform Crowdcube—250 percent over the target!

"We showed that we can make energy fun," said Kemball-Cook. "The way I see it, you can't hug a wind turbine or a solar panel. But the act of taking in energy [by] engaging people is just like hugging a floor. So we've really made it tactile so that people can see, touch, feel, and be part of the city around them and make a real difference."[4]

Pavegen had gamified energy production.

DON'T BE A SQUARE

Kemball-Cook could have stopped there. In fact, he did consider limiting his focus to big events and energy awareness. Eventually, however, he decided to pursue the grander vision of making his invention part of the world's clean energy mix.

The basic technology behind the Pavegen tiles is nothing new. For years, kinetic energy recovery systems have been used in race cars, buses, and consumer vehicles like the Toyota Prius. In these cases, the kinetic energy produced from hitting the brakes is converted into electricity that powers the vehicles' headlights, taillights, spark plugs, and so on. With Pavegen tiles, kinetic energy is produced when you step on the device, which depresses the surface by up to a centimeter. (Kemball-Cook compares the sensation to walking on a children's soft-surface playground.) This downward force drives an energy-storing flywheel inside the tile, which spins to convert kinetic energy into electrical energy through electromagnetic induction.[5] The floor tiles can be placed on any surface where people are likely to tread—from city sidewalks and train stations to airport terminals, art museums, and football fields.

Although they were impressive, early iterations of the tiles suffered from several problems, including low energy production and the fact that people didn't always step on the portions of the tile where the generators were located. To solve both issues, Kemball-Cook came up with another radical idea. He designed a *triangular* tile. This allows for a generator in each corner, which more effectively captures energy, no matter where on the tile you step.

The triangular tile (called the V3) generates five continuous watts of power, making it two hundred times more efficient than the previous design. Although switching from a square tile to a triangular one may not seem radical, consider how many triangular floor tiles you've seen in your life. Any? A triangle shape violates the conventional template for paving tiles.

Pavegen also conducted tests to improve the tile's resilience. As Kemball-Cook explained, "Launching a new technology that operates in urban environments was always going to be challenging. Unlike an app, we're designing and building a complex physical product that must operate reliably in all conditions. City streets are constantly undergoing challenges, from extreme temperature variations to a wide range of forces and impacts. Engineering this versatility into our system has been a big challenge, and it has been a highly iterative process to get to where our design is today."[6]

CONTINUOUS EXPERIMENTATION

To stress-test quicker, Pavegen decided to install the tiles in a variety of harsh environments—cold, wet, or hot (on sites as far afield as Bulgaria, Nigeria, and Brazil). To test for durability, tiles were installed at the Paris marathon finish line and in the underground station feeding spectators to the Olympic Park in London.

The company's most recent pivot was based on the realization that the tiles could generate not only light but also data. Adding sensors to the tiles can deliver information on how people move through cities. "You can use it to control lighting more efficiently," said Kemball-Cook. "It's also a really key way for retailers to know how many people are visiting their shops. We imagine Google will cover streets with this in the future and use the data in interesting ways."[7] Revenues from data generation may also provide the company with an additional way to subsidize its ongoing experiments to improve productivity.

In the near term, however, this breakthrough product is unlikely to replace wind or solar energy. It doesn't generate enough electricity to power major appliances, much less homes and businesses. However, it *does* offer a complementary source of power—one that can provide on-demand lighting in high-foot-traffic areas such as travel hubs and busy retail sites. More significantly, it also opens up the possibility of harvesting other sources of kinetic energy. Pavegen currently has "a suite of 10 patents around the world," said Kemball-Cook. "Anything that moves, we can generate power from it. Think flooring, buildings, roads."[8]

WHAT IS EXPERIMENTATION?

Experimentation is the process of turning a promising idea into a workable solution that addresses a real need. The top reason start-ups fail, according to start-up founders, is that they offer something nobody wants.[9] Therefore, to establish whether an idea is desirable and viable, you must engage in experimentation. Just as important, experimentation is an essential tool for exploring your options and testing your assumptions.

When executing an idea, you can't be driven purely by gut instinct. Through action, you must gather evidence that not only *can* your idea be built, but it *should* be built. Kemball-Cook's experiments were designed to find ways of maximizing the energy captured from each footstep *and* to prove that there would be a demand for his product if he built it.

His experiments were also meant to explore alternative designs, potential uses, and new opportunities for the product. In fact, a distinctive feature of ALIEN thinkers, as they experiment, is their ability to combine two seemingly contradictory qualities: total focus (commitment) and total flexibility.

By contrast, most conventional experimenters are single-mindedly focused on validating a hypothesis. A scientist, journalist, or entrepreneur develops a hypothesis and then sets out to prove (or disprove) the hypothesis through experimentation. The emphasis on validation is rooted in the scientific method, which is held up as the ideal of experimentation. Valid scientific protocols are based on formulating hypotheses, testing them (with treatment and control groups), analyzing the results, and drawing valid conclusions. Although these protocols can help you move away from gut instincts in search of valid evidence, too much focus on validation can stifle flexibility and open-mindedness.

In the domain of innovation, the lean-start methodology epitomizes hypothesis-driven experimentation with its mantra of "test and learn."[10] You validate individual hypotheses as you go along.

In organizations, a commitment to experimentation and data collection helps combat the tendency for idea selection and product development choices to be determined by the "HiPPO" (highest-paid person's opinion). It lets you proceed in a more deliberate and systematic way to discover the truth of what works—or, more precisely, what works when.[11]

There are many ways of testing your concepts, with more or less realistic prototypes, many of them relying on some form of deception—in line with the recommendation originally coined by Alberto Savoia: "Fake it before you make it."[12]

The Belgian-based consultancy Board of Innovation lists more than twenty experiments you can run to validate your ideas.[13] One of the most famous is the Wizard of Oz test, which manually simulates the response of an automated system. This form of testing was used to develop early speech-recognition software. Because the original recognition algorithms were slow and unreliable, the designers created a usability test in which participants spoke into a microphone while an expert typist, listening on headphones in another room, transcribed what they said on a computer. This allowed the designers to observe user responses on the computer screen while they continued to work out the system's bugs.[14] Other experiments include smoke tests (using emails or a landing page to test the demand for your value proposition); explainer videos (to show how a service would work); and app mock-ups. None of the experiments requires a working prototype.

Regardless of the methodology, the purpose of experimentation is always to test an assumption and learn from it, so you can decide whether to persevere, pivot, or pull out. Unfortunately, the traditional emphasis on validation often crowds out investigation. This approach is geared to removing uncertainty, not to identifying possibility. As a result, conventional innovators are liable to miss out on unexpected insights. They mostly test to confirm their hypotheses and assumptions rather than testing to *learn*.

PITFALLS AND FALSE EXPERIMENTS

When your aim is to validate a hypothesis, you are unconsciously drawn into a logic of supporting what you already believe to be true, rather than seeking objective answers. You look for ways to bolster existing assumptions instead of exploring uncharted paths. You become prone to confirmation biases—in terms of selective attention to data and skewed interpretation of the data. You see what you expect to see. You're blind to unexpected data. You stop looking for unanticipated findings.

As a result, you can easily become a prisoner of your initial assumptions, as shown in research on the "imprinting effects" of early decisions during a company's creation, including the type of people hired. These decisions set a trajectory that dictates subsequent moves and becomes difficult to alter.[15] Adhering to initial assumptions may cause you to be insensitive to any feedback that doesn't fit them.

Such an approach might have encouraged Kemball-Cook to settle for running a "fun company" (there was a possible product-market fit). Had this happened, Pavegen might have focused on staging big one-off events and working with schools and museums to raise awareness of energy efficiency instead of pursuing a credible clean-tech solution.

The tiles would have remained novelty products.

The Segway Trap

When innovators fail to challenge their initial assumptions, the consequences can be dire. A case in point is the Segway.

Dean Kamen, inventor of this motorized, self-balancing personal vehicle, predicted it would "be to the car what the car was to the horse and buggy."[16] He believed Segways would be used everywhere: from theme parks to battlefields to factory floors, and in cities from Seattle to Shanghai. "Cars are great for going long distances . . . but it makes no sense at all for people in cities to use a 4,000-lb. piece of metal to haul their 150-lb. asses around town," he said.[17] Kamen envisioned urban centers freed of cars to make room for millions of Segway-powered pedestrians.

In 2001, *Time* magazine described the Segway as "a machine that may be the most eagerly awaited and wildly . . . hyped high-tech product since the Apple Macintosh."[18] Steve Jobs actually stated that it might be bigger than the personal computer.

The Segway's self-balancing system was indeed a technological marvel, and Kamen—with several breakthrough inventions under his belt—had no trouble attracting funding.

Unfortunately, the entire project was so shrouded in secrecy that the Segway went through three or four iterations based exclusively on internal feedback, rather than user feedback. Concerned that someone would steal his idea, Kamen didn't seek out enough critical customers.[19] The few outsiders who *did* try the full prototype, including Steve Jobs and Jeff Bezos, were seduced by its novelty and keen to invest in it.

But while Jobs and Bezos were impressed with the vehicle, both also raised objections. Jobs criticized its design (its shape and feel) as too traditional, and Bezos foresaw regulatory concerns about using the vehicle on sidewalks. But by the time they made these points, the Segway was too close to launch for radical changes. And neither of these tech titans asked hard questions about its utility. Setting aside the wow factor, the Segway had little to offer over traditional alternatives.

Compared to a bike, the Segway was better at climbing hills, but it was also slower, heavier, and had a shorter range. In addition, it couldn't carry groceries, couldn't be ridden on roads, was much more expensive, was harder to repair, and provided no physical exercise for the user. It was "greener" than a moped, but it was much slower and couldn't carry cargo or a passenger.

Amazingly, the only person who seriously questioned its function was Aileen Lee, a young associate of the chief investor. At one meeting, she wanted to know its "value proposition," as well as its biggest competitive advantages over alternative products. The head of marketing reportedly considered her question a "cheap shot to make her look good."[20] Even Jobs's comments were not relayed to the designers on the grounds that "he didn't say anything specifically helpful and he's not even an investor."[21]

Kamen predicted that the Segway would become the primary mode of transport within a decade of its launch and force the redesign of cities. Today, nearly twenty years later, it remains a niche product, most notably used by police officers on patrol, tourists on guided tours in big cities, and mall security guards. Although Segway technology inspired the creation of a hands-free electric wheelchair, steered by leaning in the direction you want to go, this is hardly the mass market product that Kamen envisioned.[22]

The case of the Segway illustrates the drawbacks of persisting too long without critical input from customers (internal iteration can take you only so far) and the difficulties of changing course as commitment grows. Also, Kamen's personality and record as an innovator meant that his opinion carried excessive weight in internal discussions (HiPPO). Moreover, because financing was not an issue, Kamen didn't have to deal with any pushback to secure additional money. He was allowed to believe that he'd invented a great product until it was too late—until market realities shattered the grand delusion.

Though an ALIEN thinker in many respects, Kamen neglected the back-and-forth needed to determine what potential users really wanted. Without extensive external tests, he didn't gather enough information to redesign the vehicle to meet consumers' wants and needs. It was a case of focusing too much on the invention and missing the real opportunity for a breakthrough innovation. Kamen was so passionate about his idea that he failed to question his assumptions, bypassing the chance to learn about other points of view or alternative solutions.

Just because you engage in testing doesn't mean you are being rigorous. The chief risk, once you start testing your cherished idea, is that well-established confirmation biases and sunk-cost effects will deaden your responsiveness to corrective feedback. You may set up experiments to ratify your assumptions rather than refute them. You may engage in false experiments with unreliable feedback providers—family, friends, or like-minded thinkers—rather than real users. And the more you invest, the harder it may be to abandon the project or refocus your efforts.

In the wake of harsh feedback from Steve Jobs about the Segway's design, the response was "that's just Steve." Instead of considering Jobs's negative feedback, Kamen chose to focus only on his positive comments—for example, his characterization of the technology as "amazing."[23]

Talking about experimenting to *validate* versus experimenting to *investigate*, Bertrand Piccard offered an analogy. He'd noticed that wasps and bees often came into his house, but only the bees failed to escape. They would keep banging against different parts of the same windowpane in their search for a way out, whereas the wasps would move across windows.[24]

When you experiment to validate, you think like a bee. You have the illusion of testing, iterating, and reiterating, but every test occurs within the same frame. You need to experiment like the wasp: persist in your efforts, but conduct your experiments on different frames. One of those frames may provide an unanticipated opening.

ALIEN THINKER EXPERIMENTATION

ALIEN thinkers proceed differently. They use experiments to validate *and* investigate. Investigation is a more open-ended form of experimentation. It is designed to illuminate overlooked factors or preferences. It leaves room for unexpected insights to emerge.

Experimenting is not meant to lock you into a cycle. It provides you with an opportunity to go in very different directions. ALIEN thinkers remain open to sharp changes in direction. They test their offerings earlier, on a smaller scale, and in more hostile environments (like Pavegen). They encourage criticism, and they take it on board. Even when they are in validation mode, ALIEN thinkers preserve space for discovery. They test to *improve*, not just to *prove*.

To preserve your capacity to investigate (make new discoveries and capture opportunities), even as you try to validate assumptions, you must *welcome* and *accept* surprises.

Welcome Surprises

ALIEN thinkers look for ways to acquire faster, richer, and more unexpected data from the outside world. A prime example of such an ALIEN thinker is architect Frank Gehry, whose pioneering and award-winning designs are the product of a very unorthodox approach to experimentation.

Gehry's fame stems from the instantly recognizable aesthetic of his buildings. His architectural sensibility was immortalized in an episode of *The Simpsons* in which Marge writes a letter asking him to build an opera house for her hometown of Springfield.[25] Gehry balls up the letter and tosses it to the ground. But after looking at the crumpled letter, he's suddenly inspired to build a similar-looking concert hall for Springfield.

A Gehry building starts with a distinctive sketch. As captured in Sydney Pollack's classic documentary *Sketches of Frank Gehry*, Gehry's style is fluid but messy. His initial sketches are unconstrained by architectural norms. They are loose, squiggly, and impressionistic rather than precise—a far cry from the clean lines of conventional architectural renderings. Gehry's sketches pay no heed to practicalities or even gravity, but this impulsive quality prevents him from becoming formulaic or repetitive in his designs. Although the drawings provide only loose directions, in retrospect, you can see that they contain the dynamic energy of the final design. They are intended to convey a feeling that will excite—or sometimes disturb—the client.

The journey from rough sketch to finished building is a long one, and though Gehry quickly translates the sketches into three-dimensional models, there is a prolonged back-and-forth between models and drawings. In other words, the experimentation process remains in a "liquid state" for a long time. One reason for this is that Gehry resists the temptation to settle on the "right" model, preferring instead to create a variety of study models—what he dubs "shrek" (a Yiddish word for "fright") proposals.

Gehry wants the shrek models to make clients nervous. His aim is to provoke a strong reaction, whether positive or negative, so he can better understand how the client thinks. Based on the client input, he makes refinements and creates new models. These successive models don't build on the ones that came before. Instead, they represent divergent solutions. This process enables Gehry to continue exploring and learning from the clients' discomfort. In this way, his ideas gradually ferment and mature over time.[26]

Though he favors unorthodox forms, Gehry is very mindful of function. When the Walt Disney Concert Hall opened in 2003, it was widely praised for its acoustics. Esa-Pekka Salonen, the music director of the Los Angeles Philharmonic, who worked closely with Gehry on the design, commented: "Frank was very clear from the beginning. He said, 'This is a hall for the orchestra, and this is a building for music. And that has to be the first priority, and everything else is of lesser importance.' And I thought this was quite a statement from an architect."[27]

Because his designs are unconventional, Gehry is sometimes misrepresented as someone who imposes designs on clients—take it or leave it. "That's absolutely the opposite of who he is," said *Vanity Fair's* architecture critic, Paul Goldberger. "He goes through multiple iterations of every project and it's not only his thinking but he's very, very eager for feedback from clients and for dialog with them."[28]

Propose Multiple Models

Gehry proposes a multitude of designs. This gives him, as an experimenter, two advantages.

First, it keeps him from getting too attached to any one idea. You can more easily remain open to learning—and pivoting—if you are not overinvested in your prototype. The longer you pursue an idea, the more you feel as though your reputation is tied to it. As Gehry's colleague, Jim Glymph, put it: "If you freeze an idea too quickly, you fall in love with it. . . . If you refine it too quickly, you become attached to it. And it becomes very hard to keep exploring, to keep looking for better. So the crudeness of the early models, in particular, is very deliberate. . . . It keeps you from bonding with the idea so much that you can't move on."[29]

Second, it speeds discovery and lets you pivot earlier, before hitting a wall. The lean start-up approach is very much sequential. You set off in one direction, go through iterative cycles of build-measure-learn, pivot when an assumption is clearly refuted, and progressively home in on a product-market fit. By contrast, ALIEN thinkers move forward in parallel. They create multiple concepts and pursue the option that seems to have the most traction with future users, as shown by page views, sign-ups, or presales.

The primary aim is to get feedback and a first impression of whether the proposal resonates with independent experts or actual paying customers. Market testing in parallel is all the easier in the digital realm, where it's fast and cheap to realistically simulate nonexistent products or services through landing pages. This way, no opportunity is lost or inadvertently dismissed.

Of course, anyone can propose multiple models, but how do you trigger the kinds of reactions that ALIEN thinkers like Gehry find so valuable?

Provoke Extreme Reactions

Gehry doesn't just propose several sketches or models. He develops extreme points of view that provoke strong emotional reactions. Why? Because ALIEN thinkers value negative feedback as much as positive. This is why Gehry uses the shrek models. It's about testing people at the boundaries, making them ill at ease so they will react.

One professor involved in the committee overseeing the construction of the Peter B. Lewis Building at Case Western University recalled the experience of working with Frank Gehry: "He often would say about a model he was presenting to the university team: *This is not what we are doing*. And it was difficult to appreciate what he meant until we followed the design as it evolved through dozens of iterations."[30]

The crude sketch of the early model was just a means of getting a reaction, exploring possibilities, and investigating issues relevant to the building project. Though his fluid sketches are short on detail, they capture the energy and spirit of the proposed vision. And because his alternative models don't build on each other, as if they were presented in sequence, Frank Gehry is able to constantly surprise clients and maintain their discomfort to encourage a steady stream of feedback throughout the design process. Gehry is not interested in knowing what the building will look like at the very start. He's willing to engage in open and vibrant discussions with clients so that together they can find out.

Gehry's approach highlights the importance of generating prototypes that will provoke honest feedback. For example, Pavegen elicited quality reactions to its early concept ideas by trialing a working installation at Kemball-Cook's old school, filming how students interacted with the tiles and how the tiles stood up to the treatment. When the kids went crazy for them, Kemball-Cook realized that he had stumbled on the gamification of energy production.

You can also test your concepts in extreme environments, as Pavegen did in Nigeria and Brazil, to generate accelerated feedback and to iterate faster.

Beyond the problem of generating slow and insipid feedback, there's also the challenge of digesting feedback. Even if you ask the right questions and receive valuable answers, you may fail to properly integrate them into your thinking.

Accept Surprises

ALIEN thinkers ask the right questions and *heed* the answers. They *act* on feedback.

Eliciting rich data is one thing, but once you get those inputs, how do you make sense of them? And are you prepared to deal with the implications? Even if you acquire robust feedback through parallel testing or provocation, you could fail if you aren't receptive enough to the findings.

We know from the Segway case that a passion for your idea can get in the way of hearing constructive and well-intentioned feedback. So how can entrepreneurs stay objective? After all, the passion required for entrepreneurship often stems from your love affair with your vision or creation. You must not only invite feedback; you must also accept and leverage it.

Consider the case of Wysa, the brainchild of husband-and-wife team Ramakant Vempati and Jo Aggarwal. Because Vempati, a former Goldman Sachs banker, and Aggarwal, an international managing director at Pearson Learning, worked far from their aging parents and relatives, they were seeking a convenient and effective way to care for them.

Leveraging their joint knowledge of product innovation, business analytics, and e-learning, they hit on the idea of using technology to facilitate remote care. They thought they could use passive sensing to track factors that would help detect possible signs of depression.

Initially they considered making a wearable device, but soon they realized that they could collect the necessary information without the need for a new gadget. A smartphone app could capture data (with user permission) about changes in calling patterns and sleeping patterns, as well as whether someone is moving around more or less frequently—and more. As it turns out, once such data is aggregated, it's a very reliable indicator of depression. In fact, it's capable of identifying sufferers of depression with 85 to 90 percent accuracy.

There was just one problem. "As we did our trials, we found that people weren't ready to see a therapist," said Aggarwal.[31] People don't always want their loved ones or doctors to know if they're suffering from depression or anxiety.

The founders now realized that they had created a kind of fire alarm. What was missing, however, was a fire engine—a way to respond to the alarm once it was raised. So they pivoted. They started testing a chatbot, combining AI with passive sensing to help people in real time. That's how Wysa was born.

Personified by a chubby penguin, Wysa is a virtual therapist designed to give people access to techniques that help them build mental resilience and cope with daily stress. It was developed with the help of a therapist, who reviews all content. It had an initial test audience of around fifty, including the Wysa team and the company therapist's network of psychologists.

When it was rolled out in 2017, Wysa was an immediate hit. "It started off as a little experiment and it's completely gone viral," said Vempati. "We don't market it, there is no social media campaign, there [are] no Google adverts, Facebook marketing. . . . Nothing. People are finding it because they are looking for something like this."[32]

"A lot of people are not willing to talk to another human being because they fear dependence, they fear judgment, they fear weakness, but they're willing to talk to a bot, and if that bot can appear compassionate, which Wysa is designed to do, that can make all the difference," said Aggarwal.[33]

Today, Wysa is helping not only people who refuse to see a therapist but also those who have no access to one. India, for example, has just five thousand mental health professionals for 1.3 billion people. The founders of Wysa also noted that there is a huge correlation between mental health and physical health. Depression is the number one predictor of nonadherence to medical regimens.

And the product keeps improving.

"Initially four to five percent of Wysa's users used to say things like, 'You're [not] getting me,' 'You're not understanding me,' and now 0.4 percent of Wysa's conversations end in that kind of stop," said Aggarwal. "And all we've done in the last year is figure out how to reduce that number. The users tell us when it's working, when it's not working. And we figure out areas which are not working, and we keep fixing them and giving them alternatives."[34]

During their journey, Wysa's founders discovered that empathy is not merely a trait; it's also something you can deconstruct. The most empathetic responses that Wysa can offer, they have learned, are not statements but questions. By asking how someone is feeling, acknowledging those feelings without trying to change them, and then asking additional open-ended questions, the bot encourages people to express their feelings and helps them feel less alone.

Although Vempati and Aggarwal don't tout Wysa as a replacement for therapists or a cure for depression—they bill Wysa as "your 4 a.m. friend for when you have no one to talk to"—there are signs that it helps reduce depression by up to 30 percent over the course of two to four weeks. Growing entirely by word of mouth, it claims over one million users in thirty countries and has a user rating of more than 4.8 out of 5, so it appears to be doing something right.

Ultimately, the founders are hoping to integrate Wysa into routine health maintenance regimens. "If we can build psychosocial support and depression support into care as usual, for a physical health condition, we feel that we would be able to really make the physical health better and get to a lot of people who wouldn't otherwise classify themselves as needing mental health support, especially in countries like India where people will not acknowledge that they need help," said Aggarwal.[35]

After a couple of pivots, what had started as a wearable device designed to assist with family care became a smartphone app that may someday become an integral part of generalized health care.

The Wysa story illustrates two key facets of accepting feedback. Whether you're working alone or in a team, you need to think about how you interpret incoming data and how to avoid biases. There are two ways to address these challenges: (1) let the data speak, and (2) seek out people who think differently than you.

Let the Data Speak

Vempati and Aggarwal discovered that the right mix of passive sensors produced robust warning signs that someone was falling into depression. They could have stopped there. They had invented a breakthrough system for identifying people at risk and alerting their relatives. But one trial revealed the weakness of the solution: thirty participants were flagged by the app as potentially having clinical depression and were advised by the supervising doctor to seek therapy, but only one person followed the doctor's recommendation. The duo could have discounted this finding as someone else's problem, but they didn't. Instead, they recognized that they'd invented a fire alarm without providing a fire engine, so they pivoted to a chatbot.

Unfortunately, not every entrepreneur, innovator, or artist is as receptive to feedback as the Wysa team.

As cofounder of Pixar (and later president of both Walt Disney and Pixar Animation Studios), Ed Catmull learned this the hard way. At Pixar, Catmull oversaw a parade of film directors who were trying to bring their creative projects to the screen. During this time, he noticed systematic patterns of resistance to constructive and well-intentioned feedback. At some point in the life of a project, the director's passion would blind them to the movie's inevitable problems.[36]

In his book *Creativity, Inc.,* Catmull recalled directors' reactions when receiving feedback from senior colleagues. When the stakes are high and people in the room don't seem to understand a director's vision, he said, "it can feel to that director like everything they've worked so hard on is in jeopardy, under attack. Their brains go into overdrive, reading all of the subtexts and fighting off the perceived threats to what they've built."[37]

Although the feedback might pinpoint what's wrong, missing, unclear, or illogical, candid, incisive, and empathic feedback is only valuable if "the person on the receiving end was open to it and willing, if necessary, to let go of the things that don't work."[38] Catmull stressed that the best feedback "can't help people who refuse to hear criticism without getting defensive, or who don't have the talent to digest feedback, reset, and start again."[39]

By contrast, one key to Frank Gehry's success is his ability to admit failure—to reject his initial attempt (in part or altogether) and start over with renewed confidence that the next draft would be better because of what he'd just learned. With one wealthy client, Peter Lewis, he worked for twelve years in order to get the project right—with the client only too happy to share in the adventure and foot the bill. It's an extreme illustration of maintaining absolute focus and absolute flexibility in parallel.

Catmull advised those giving feedback to be clear that the project, not the creator, is under the microscope. This is a critical principle: "You are not your idea, and if you identify too closely with your ideas, you will take offense when they are challenged."[40]

It's not just about gathering feedback but also processing it honestly, without letting your ego, biases, or attachment to the idea get in the way. You need enough resilience to accept criticism, learn from it, and move on.

Letting the data speak means interpreting it without prejudice. A complementary tactic is to invite alternative prejudices that will challenge your thinking.

Seek Out People Who Think Differently

Another key to accepting critical feedback is to discuss it with people who may see other possibilities in the data and won't be scared to challenge you. In other words, seek out people who think differently, including those whose views tend to be the opposite of yours.

In the case of Wysa, the complementary perspectives, skill sets, and professional experiences of the two founders made it easier for them to envisage further opportunities and to pivot, instead of stopping, after they found their first product-market fit. In the case of Pavegen, Laurence Kemball-Cook had his father—not an engineer but a former management accountant, with no experience in start-ups but a long career in big business—as chairman. Such relationships help you combat the tendency to fall in love with your own ideas.

If you don't have a cofounder or trusted partner who will offer different interpretations of test results, look for someone who can fill that role. The people you consult might be current team members or outsiders—anyone who can help you expand your perspective.

The idea of bringing together people with dissonant views is familiar from the imagination phase (see the section on the "Medici effect" in Chapter 4). However, the capacity to engage in divergent thinking is also relevant during experimentation. Together, you and your devil's advocates (opposite thinkers) may not capture the *full* picture of reality, but at the very least, gathering more "pixels" should prove very useful.

To combat filmmakers' resistance to feedback at Pixar, Ed Catmull created a group called the Braintrust. It consisted of a panel of experts (from a core group of directors, writers, and heads of story) who delivered candid feedback to help directors pivot when necessary. In addition to the quality of the feedback, the key to the Braintrust's impact was its two ground rules: (1) no hierarchy, and (2) no obligation to implement the feedback. As a result of these rules, directors didn't enter discussions in a defensive posture and were less prone to emotional reactions. Therefore, they were more likely to hear, digest, and act on the feedback.[41]

You don't have to work at Pixar to create a Braintrust. As an innovator, you can surround yourself with people capable of challenging your interpretations of the evidence. Catmull saw this as the best antidote to confirmation biases: "Seek out people who are willing to level with you, and when you find them, hold them close."[42]

ALIEN thinkers use experimentation differently than most businesses do today. They conduct experiments to expand their knowledge through *investigation*, not just to sharpen their knowledge through validation. And digital technology pushes this logic even further.

DIGITAL EXPERIMENTATION: WHO NEEDS A HYPOTHESIS?

ALIEN thinkers are less concerned with conducting perfect experiments than making better decisions. To this end, they use digital tools to accelerate experimentation and achieve three goals that, until recently, would have seemed contradictory: trial without error, feedback without moving, and testing without hypotheses.

Trial Without Error

Of course, there will always be errors, but when experiments are powered by digital technologies, the cost of the errors—in terms of time, money, or physical damage—are greatly reduced. With computer simulations, you do not endanger the brand by interacting with the real world. By interacting with a simulated world, you protect your company and its products and services from unnecessary risk while gleaning valuable insights.

For example, to test his *Solar Impulse* model, Bertrand Piccard (see Chapter 3) created a "digital twin" of the plane, using 3D software to design and test individual parts and complex assemblies. This allowed his team to forgo slow and costly physical prototypes and simulate the plane's performance under a variety of conditions, slashing the need for numerous dead-end experiments.

NASA has been using a form of digital twin since *Apollo 13*. Also known as "mirrored systems," they allowed engineers on the ground to create a fix for any problems with the spacecraft and then transmit the solution to the astronauts in space. The difference today is that the mirrored systems are fully digital. NASA now creates digital twins for all its systems, vehicles, and aircraft. Indeed, information collected from sensors on the physical equipment is fed into the digital twin in real time, so that it can evaluate the equipment after it's actually built.

The principle of using digital twins for modeling is not new, but advances in computing mean that the research and development can be done at a very low cost and a high speed, without the risks associated with traditional approaches.

"The ultimate vision for the digital twin is to create, test and build our equipment in a virtual environment," said John Vickers, manager of NASA's National Center for Advanced Manufacturing. "Only when we get it to where it performs to our requirements do we physically manufacture it. We then want that physical build to tie back to its digital twin through sensors so that the digital twin contains all the information that we could have by inspecting the physical build."[43]

NASA has even created a digital twin of its entire Langley Research Center, a 764-acre campus where the Apollo lunar module was designed. The digital twin is used to optimize the facility's operations, including its airport and forty wind tunnels, as well as to ensure the safety of its three thousand personnel.[44]

Feedback Without Moving

The lean start-up methodology emphasizes "getting out of the building" to elicit feedback on concepts, insights, or value propositions. Today, this same information is available through remote usability testing and live prototyping.

For example, software produced by companies such as Optimizely, Unbound, IDEO, and Usabilla allow for automated A/B testing of websites.[45] The software makes a small adjustment to a website, such as changing the size of an image or moving text around a page, and then tests the performance of the site with the change (the B version) against the original version (the A version). When conducted at scale, the process can result in sites that load faster, gain more traffic, and attain a higher degree of stickiness. This kind of simulated process of natural selection can be done without any direct human intervention.

COVID-19 tracing apps provide an interesting example of collecting massive amounts of data digitally in a situation where it may be unsafe or unethical to collect it physically. At the time of writing, governments around the world are experimenting with contact tracing and communication tools to identify and inform people who may have been infected with the virus.

Testing Without Hypotheses

The idea of testing without hypotheses challenges one of the fundamental tenets of the scientific method. Keep in mind, however, that a hypothesis is merely an educated guess, typically based on a theory. Given the power of computing today, we no longer need to make guesses. Instead, we can continually test everything!

In the online world, you don't even need to simulate. Live and continuous A/B testing allows for perpetual investigative experimentation.

In A/B testing, there doesn't even need to be a reason for the test. No one needs to hypothesize, "If we make the text larger, more people will click on the link." You can simply test the link in different formats to learn which one gets the best result. The idea is not to *prove*, but to *improve*. In fact, A/B testing moves much of the burden from the experimenter to the user. It is the *consumer*, not the designer, who decides which design is the best.

The ALIEN experimenter doesn't need to formulate a hypothesis—just come up with an experiment and then measure the results. He or she doesn't need theory—just the openness to conceive different outcomes, as well as the ability to measure and learn from them. In short, the experimentation process is turned on its head—from "ready, aim, fire" to "fire, ready, aim."

The era of the scientific method has come and gone. It was critically important when evidence was the bottleneck. Because evidence gathering was expensive, time consuming, and imprecise, hypotheses were paramount. That is no longer the case. Evidence gathering is cheap, fast, and accurate, so hypotheses are less important.

One example is the release of SimCity 5, one of the most popular video games created by Electronic Arts (EA). It sold over a million copies in the two weeks after its launch in 2013, but it didn't initially look like it would achieve this goal. EA released a promotional offer to drive more game preorders on its website. The offer was displayed as a banner across the top of the preorder page, but the results were disappointing, so they decided to try some different approaches.

The test produced some surprising results. When the promotional offer didn't appear on the preorder page at all, sales increased 43.4 percent. As it turned out, people just wanted to buy the game. They didn't need any extra incentive. Most EA staff thought the promotions would increase purchases, but this turned out to be totally false.[46]

Another example comes from Google's AI group DeepMind. DeepMind works with the Cancer Research UK Imperial Centre at Imperial College London to improve the detection of breast cancer. The mammogram scans used today miss thousands of cancers per year and/or lead to false alarms, so researchers are using historical data (from more than 7,500 patients) to test whether machine learning can detect more cancers, more accurately, from the mammograms than doctors. DeepMind employs a "brute force" analytical approach, meaning that it tests every possible outcome rather than making the educated guesses that form the basis of traditional hypothesis testing. It's hoped that if machine learning outperforms the doctors, it can minimize errors and improve breast cancer treatment.[47]

Ethical Considerations

For better or worse, digital tools also make it easier to deceive people. It is increasingly common to propose and then test a product that doesn't yet exist (and may never exist). For example, Laurence Kemball-Cook of Pavegen engaged in testing that was, strictly speaking, illegal by entering a building site and installing tiles there without permission, and then talked about it as if it were a real installation. Although terms such as *growth hacking* and *guerrilla marketing* give such practices a veneer of legitimacy, this kind of deceit can have serious repercussions.

One now-famous example is Theranos, which proposed to perform a revolutionary range of medical diagnostic tests on the tiniest blood sample. The company claimed to have developed a proprietary blood analysis machine called Edison, but it was actually running the blood tests on a series of standard devices purchased from other companies, including Siemens.[48] The interest was huge, and in 2012 Theranos, under the leadership of Elizabeth Holmes, received massive investments from national pharmacy and supermarket chains Walgreens and Safeway to bring the technology within easy reach of consumers. And yet the Edison device continued to produce inaccurate results. By the time the fakery was fully exposed in 2016, Theranos had already opened forty "wellness centers" inside Walgreens stores. Unfortunately for Theranos and Holmes, simulation that is acceptable at the experimentation stage is considered fraudulent once the business is launched.

Theranos is the most infamous case of such dishonesty, but there are others. For example, the start-up Magic Leap raised hundreds of millions of dollars in 2016 on the strength of videos "demonstrating" its augmented reality (AR) headset, which promised stunning 3D visuals that would combine elements of its virtual world with the real world. Although the company insisted that its AR was built with a brand-new imaging technology, it would not divulge any information about the tech, and it later emerged that the demo videos were created using special effects.

Then there's The Doom That Came to Atlantic City. To obtain financing for this Monopoly-style board game, the inventor, Lee Moyer, launched a Kickstarter campaign that blew past its original $35,000 goal, attracting nearly $123,000 in pledges by the time the campaign ended in June 2012. Fourteen months later, Moyer told his Kickstarter backers that he would refund their money because the project was being canceled. But he never refunded a penny. Instead, according to the US Federal Trade Commission, he spent the money on himself.[49]

To combat such transgressions, companies such as Intuit are introducing rules for ethical experimenting.[50]

PUTTING ALIEN THINKING TO WORK
Testing Times for SNCF

In 2014, French rail giant SNCF, famous for developing the fastest-ever train, was facing a huge challenge. It needed to transform itself into an integrated group. And to stay competitive, it had to broaden its focus beyond trains and technological innovation.

French rail services were not in a good place. Two high-profile incidents had exposed dysfunctional relations between the state-controlled railway operator (SNCF) and the infrastructure owner (RFF). In 2013, an intercity train derailed as a result of poor maintenance on the tracks and faulty collaboration protocols among staff of the two state-run entities. The following year, a fleet of new trains ordered by SNCF turned out to be too wide for many of the stations they were due to serve. RFF had sent over the correct measurements—but only for stations built in the last thirty years.

The French government instigated a reform to bring the two entities under one umbrella to promote information exchange and cooperation between them.

Their respective presidents turned to us for help. Their immediate preoccupation was how to improve collaboration for operational reasons: to conduct planned maintenance and regenerate an aging infrastructure without disrupting services or compromising safety—and, of course, to avoid past mistakes.

But when we interviewed the top teams of both entities, another pressing challenge emerged. The newly formed group needed to redefine its commercial ambition and service offerings. The competitive arena had shifted. Perceived as too expensive, SNCF was losing passengers to carpooling, intercity buses, and low-cost flights. Users expected cheaper and more flexible travel options from SNCF. They expected additional services. It was no longer just about well-designed, reliable trains or pure speed.

To respond, the group needed to gain in agility and entrepreneurial spirit. Both SNCF and RFF were state entities with a strong culture

of top-down leadership and adherence to written rules and proce-
dures. This limited the capacity of local decision-makers to pursue
initiatives on their own. As one of them pointed out, "A great idea that
makes sense right now could look quite ridiculous if it takes us three
years to make it happen."[51]

The long innovation cycles and rigorous risk assessments that
characterized big technological advances at SNCF were not suited
to innovation in other domains. To improve organizational processes,
respond rapidly to new market opportunities or threats, and come up
with commercial and service innovations, the group needed to develop new ways of quickly trialing and implementing new ideas.

Experimenting in New Ways

We proposed a learning journey to help instill a spirit of collaboration
and experimentation across the combined executive group.

The first part of the journey was virtual. We created online material to encourage executives to think differently about developing and
testing new ideas. For six weeks, participants received case studies
and articles to fuel discussions in their virtual groups about disruptive
transportation trends (e.g., ride sharing) and successful innovation in
other state-controlled organizations, such as the French national mail
service (La Poste).

With the help of the group heads of communication and strategy,
we also identified three little-known success stories within SNCF
where joint thinking and fast experimentation had produced critical
improvements. We created short films to share these with participants, which showed them pockets of innovation in their midst.

In one of the films, executives described an operational initiative
that had slashed response times to breakdowns. One emphasized the
change of mindset and collaboration required: "We went from an attitude of 'why I can't do it' or giving fifty good reasons for not doing it
to 'How can we get it done?'"[52]

Another film described the efforts of the top team in Austerlitz,
one of the main Paris stations, to rethink how its twenty-two million

annual passengers were traveling to and from their station. Even as the team discussed how to proceed, a bike-sharing concession called Vélib' sprang up in the station. One team member told her colleagues, "We should all be ashamed that a company not connected to us found a way of providing a customer service that we seem incapable of setting up ourselves."[53] This proved to be the impetus they needed to take action, converting a run-down part of the station into a transition area for two-wheelers, taxis, and rental cars. The story captured the need for SNCF to be more agile in responding to competition and its key role in making passengers' journeys more seamless.

We used these stories to stimulate discussion among the executives about perceived obstacles to rapid innovation, and we urged them to identify small problems they could solve in the field, and to imagine solutions they could test by themselves without soliciting the input and expert opinions of authorities in Paris.

To round off their journey, we gathered the group's top 650 executives for a two-day event designed to help them try what they'd learned. With everyone assembled around 110 tables in one massive space, with twenty coaches to facilitate the process, we conducted a mass "deep dive." Working in groups on one of four strategic themes, the participants had to brainstorm ideas and then give and receive feedback across groups to improve their proposals. They went through three iterations during the event. After the last iteration, the teams presented their projects and everyone voted on two criteria: whether a project was cool, and whether it was feasible. The results flashed up immediately on giant screens. The teams then had six months to continue developing and refining their proposals.

This process forced participants to set aside their inhibitions about experimentation, including their reluctance to take risks and listen to criticism. They saw how quickly and cheaply they could work collaboratively around innovation themes to generate, present, and iterate ideas about the future. They also learned to break away from their traditional ways of thinking and testing, and saw how they could tap into the wisdom of a crowd to identify ideas with traction.

Overcoming Tunnel Vision

The kinds of innovation proposed by the groups displayed a level of agility, openness, and risk-taking that belied the organization's previous reputation for inertia. There was now a clear emphasis on disruptive thinking and experimentation.

Three breakthroughs that emerged marked a sharp break with past practices and warrant special attention:

- In 2015, SNCF launched TGV Pop, a funky app inspired by Groupon. Would-be travelers vote online for a limited selection of destinations and enjoy discounted rates (provided the trains are filled). The app overturns the conventional preoccupation with encouraging passengers to book early and has proved very successful with young, hyperconnected travelers, many of whom had abandoned trains for carpooling.
- Another popular initiative is TGV Max, which drew its inspiration from the flat-rate and unlimited-access models practiced by phone companies and internet service providers. It offers young people unlimited off-peak travel on high-speed trains for a monthly fee of seventy-nine euros. It sounds trivial, but it defies the sacred principle of yield management by requiring the company to look at revenue per client instead of revenue per seat.
- SNCF has also experimented with new ventures beyond the trains themselves. In December 2015, it announced that it would be testing three new partnerships: the first with KidyGo, a start-up that connects parents with travelers who sign up (and get rated) to accompany children on train journeys; the second with OuiCar, a peer-to-peer car rental platform, to cover the last leg of the journey; and the third with Airbnb to help travelers finance their trip by renting out their empty residence.

Not surprisingly, the partnership between a state-owned company and Airbnb proved the most controversial. Although surveys showed that 80 percent of eighteen-to-thirty-four-year-olds endorsed the move, SNCF was forced to abandon the partnership after just one day because of pressure from the powerful hotel sector. Nevertheless, the failed experiment spoke volumes about the new mindset at SNCF—a readiness to envisage new models and test a tricky partnership; a commitment to letting the data speak; and a willingness to then postpone the venture when they realized it wasn't getting the hoped-for traction.

This spirit of testing to improve rather than to prove, which does not come naturally to Cartesian-minded French engineers, was captured in public comments by SNCF's head of marketing, Rachel Picard: "Taking a 'test and learn' approach enables you to be more agile, while at the same time limiting the financial risks when you launch an innovation."[54]

The new mindset is helping SNCF become an orchestrator of mobility solutions, not just a rail company.

+≫═≪+

KEY TAKEAWAYS

- Experimentation is the process of turning a promising idea into a workable solution that addresses a real need.
- The top reason start-ups fail is that they offer something that nobody wants. Therefore, to establish whether an idea is desirable and viable, you must engage in experimentation.
- As they experiment, ALIEN thinkers combine two seemingly contradictory qualities: total focus and total flexibility. By contrast, the conventional focus of most experimenters is on validation. This approach is geared to removing uncertainty, not to identifying possibility. As a result, focusing on validation often means missing out on unexpected insights.
- A commitment to experimentation and data collection helps combat the tendency for idea selection and product development choices to be determined by the "HiPPO" (highest-paid

person's opinion). It lets you proceed in a more deliberate and systematic way to discover the truth of what works—or, more precisely, what works *when.*

- The case of Segway illustrates the drawbacks of persisting too long without critical input from potential customers. Internal iteration can take you only so far.
- A big risk, once you start testing your idea, is that well-established confirmation biases and sunk-cost effects will deaden your responsiveness to corrective feedback.
- ALIEN thinkers use experiments to validate *and* investigate.
- As an ALIEN thinker, architect Frank Gehry welcomes surprises with his experiments. By creating multiple "shrek" models, he provokes strong reactions from clients that help him understand how they think. The process enables him to explore and learn from clients' discomfort, which helps his ideas ferment and mature over time.
- Whether you're working alone or in a team, you need to think about how you interpret data and how to avoid biases. There are two ways to address these challenges: (1) let the data speak, and (2) seek out people who think differently than you.

QUESTIONS TO ASK YOURSELF

1. In your organization, to what degree are product selection and development choices determined by the highest-paid people's opinions? Can you devise experiments to deliver evidence that can inform and enrich their judgment?

2. How open are you to new possibilities that emerge when trying to validate a core hypothesis? Can you recall at least one occasion when confirmation bias or sunk-cost effects impeded your ability to implement corrective feedback? What are you doing to keep this from happening again?

3. Do you have regular exposure to people who think differently from you? If not, where could you find such people?

4. Can you investigate multiple paths in parallel—like Frank Gehry—rather than prototype and test each promising idea in sequence?

5. Be honest. Do you tend to regard criticism of your ideas as a criticism of *you*? How can you reduce such defensiveness when bringing new ideas to your colleagues or boss?

On Digital

1. Can you use digital tools like a simulator, digital twin, or A/B testing to automate experimentation?

2. Can you test your ideas with remote stakeholders, through social media or online communities?

SIX

Navigation

Maneuver to Soar . . . and to Avoid Being Shot Down

WHENEVER SWISS EXPLORER SARAH MARQUIS TELLS PEOPLE about her newest expedition, she gets the same reaction: "You're crazy!" If she listened to these people, she'd cease to explore at all.

Sarah is an adventurer who's spent much of her adult life trekking solo through some of the world's most forbidding regions, from deserts to jungles to mountain ranges, living off the land and blazing her own trails. Her biggest expeditions include trekking from Santiago, Chile, to Machu Picchu in 2006; walking ten thousand miles between 2010 and 2013, including from Siberia to the Gobi Desert, then through China, Laos, and Thailand; crossing the Australian Outback in 2015; and crossing Tasmania in 2018.

Each of these big adventures has been followed by a book. Extreme hiking in unexplored environments is her passion, but sharing that experience through storytelling is her mission. She feels a responsibility to "restore our lost connection with nature before it's too late."[1] In 2014, she was named Adventurer of the Year by the prestigious National Geographic Society.

Her experiences as a woman traveling solo in remote territories underline the importance of navigation when venturing into the unknown.

ENTERING HOSTILE TERRAIN

Marquis takes a dual view of navigation, emphasizing what happens *before* as well as *during* a trek.

Successful navigation starts with preparation. To respond appropriately to threats and opportunities on the spot, she needs to be physically and mentally prepared. For a three-month expedition, Marquis typically spends two years learning about the area and planning the route and logistics, adding intense workouts during the second year.

After a preliminary visit to the region, where she talks with regional experts and visits local markets, she goes home to process the information she's acquired about the terrain, weather, and wildlife, and to determine what gear will be most appropriate. She also consults other sources of knowledge. "I am like a sponge," she said. "I buy lots of books about the place: about previous wars, the history of the land, the languages, the Indigenous peoples, the medicinal plants. And I gather data from farmers, geographers, biologists, historians. I go back to old sources because the knowledge I need is timeless. Descriptions of gold prospectors from a hundred years ago are still valid because nothing much changes in that terrain. I want to know how people died there, so that I don't die."

She also has to reach out to potential sponsors and supporters. And that means crafting an engaging story about the forthcoming journey. "Before I even start walking, I've been telling my story over and over to people who don't really care about my story or about me," she said. "Storytelling is really, really important. Can you put words on feelings, on goals, on dreams, on what matters for you? This is part of the expedition. I tell them a story that I'm ready to die for."

Many of Marquis's concerns are analogous to those of innovators trying to bring their ideas to life or challenge the status quo. For their ideas to survive, they must also anticipate the likely threats, asking themselves: What "language" do people speak in this organization or sector, and do I understand them? What historical tensions may I encounter? What gear will I need to take? Where will I find additional resources along the way? What dangers should I take into account? Whose support will be essential, and how can I convince them to support me?

Innovators often neglect this preliminary aspect of navigation. Turning a good idea into a great source of value for users takes a lot of work. As Marquis put it: "The success of a trip like that is down to the details and an understanding of your surroundings. It's not about luck. It's not about 'having a go' at something. It's a lot of preparation."

Once she starts the journey, there is a transition from planned maneuvers to dynamic navigation because she can't envisage every contingency. "You plan as much as you can, but when you take the first step, you have to be ready for the unknown. You have to read the environment and make full use of your senses—not just your sight but also your hearing, smell, taste, and touch." For example, when she's hungry and comes across an unknown plant, she may lick it, chew it, or rub it on the inside of her wrist, where the skin is thinnest, to see if it provokes a reaction and determine whether it's edible.

To navigate, Marquis doesn't rely on GPS, using it only to occasionally double-check. Instead, she favors the traditional tools of map and compass. These enable her to project herself into her environment: "My topographic maps are everything. With paper you can actually imagine the reality of the landscape. . . . With the GPS, you can't. The picture is narrow and if you move, you don't know where the north is. You have to use your brain too much."

Sleeping in a tent also enhances her feel for the terrain. It provides direct contact with the earth for prolonged periods, sharpening her sense of orientation. "You become aware of your surroundings, you can see the mountains, and you can actually feel the sun and feel the direction to go."

Overreliance on technology interferes with her awareness of her surroundings, her ability to feel, to make connections, to learn as she goes, and to adapt. On one occasion, in Mongolia, she noticed a herd of camels moving very quickly in one direction. Wisely, she did the same, and thereby escaped a coming sandstorm long before it arrived. On another occasion, after getting dehydrated in Australia, she recognized a particular type of bird that flies only about five kilometers in a day. "So if you see that bird, you know there is water nearby. You just have to figure out what direction the water's in." She is constantly on the alert. "All the time, I scan the landscape to see what it can offer me."

Like Marquis, disruptive innovators need to survive in a hostile and changing environment. They, too, need to navigate to avoid threats and seize opportunities—to learn as they go and maintain their sense of direction. Doing this requires nimble and creative responses to un-expected challenges, as well as preparation. When you have a great idea, it can be tempting to skimp on the preparatory work, especially when you're an insider who expects the organization to welcome your innovation.

WHO WANTS A DIGITAL CAMERA?

Former Eastman Kodak engineer Steven Sasson learned the impor-tance of navigation the hard way. Although his journey of exploration resulted in the first digital camera, his failure to gain management's support for the "camera of the future" illustrates how easily even a brilliant invention can be smothered in its cradle. To avoid that fate, ALIEN thinkers may need to fly under the radar while their innova-tions are in the early stages of development, recruit influential sup-porters, and position their offerings in ways that don't overtly threaten the status quo. Your success may hinge on meeting not just one or two of these challenges but all three of them.

Sasson's innovation trek began in 1973 when he joined Kodak. Soon thereafter, the twenty-four-year-old was given the seemingly minor task of seeing whether there was a practical use for a charge-coupled device (CCD), which had been invented a few years earlier. (A CCD is a sensor that detects light and converts it into digital data.)

"Hardly anybody knew I was working on this," he later told a re-porter, "because it wasn't that big of a project. It wasn't secret. It was just a project to keep me from getting into trouble doing something else."[2] As he recounted in a 2011 speech, "It was a very small project and, therefore, it had no real management, no one asking how things were going or anything like that. . . . Nobody was paying attention, we had no money, and nobody knew where we were working. So the situation was just about perfect."[3]

Over the next two years, the efforts of Sasson and a couple of technicians to discover a practical use for the CCD produced an unexpected result—a "Rube Goldberg device with a lens scavenged from a used Super-8 movie camera; a portable digital cassette recorder; 16 nickel cadmium batteries; an analog/digital converter; and several dozen circuits—all wired together on half a dozen circuit boards."[4] It was the world's first digital camera.

But it was "more than just a camera," said Sasson. "It was a photographic system to demonstrate the idea of an all-electronic camera that didn't use film and didn't use paper, and [used] no consumables at all in the capturing and display of still photographic images."[5]

Sasson and his colleagues made a series of demonstrations to Kodak executives from various departments. He brought the digital camera into conference rooms, took photos of people in the room, and uploaded the images to a television. "It only took 50 milliseconds to capture the image, but it took 23 seconds to record it to the tape. I'd pop the cassette tape out, hand it to my assistant and he put it in our playback unit. About 30 seconds later, up popped the 100 pixel by 100 pixel black and white image."[6]

Although the camera could have stored hundreds of images, Sasson deliberately set a limit of just thirty images. He chose that number because it fit snugly between the twenty-four and thirty-six images available on standard rolls of Kodak film. "I didn't want to get into a discussion about storing hundreds of images on a portable removable device," he said. "It would have complicated the story. . . . Sometimes you have to keep your idea simple and straightforward in order to keep your discussion focused on the essential elements. I set it at thirty to make it more comfortable for people to think about the idea."[7]

Sasson also opted to avoid the term *digital*. Instead, he referred to the breakthrough technology as "filmless photography."

This proved a *very* poor choice of words. In fact, it was a disastrous way to position the new technology to executives of Kodak, a company whose immense profits depended on a value chain that involved the sale of photographic film, papers, chemicals, and film and print processing. From 1975 until the early 1990s, Sasson and his team were steering a digital sailboat into hurricane-force headwinds.

Many Kodak executives were convinced that nobody would ever want to look at their pictures on a television set, noted Sasson.[8] The main objections, however, came from the marketing and business sides of the company. Kodak had a virtual monopoly on the US photography market, making money at every step of the process. Americans took photos with Kodak cameras, using Kodak film and Kodak flash cubes, and then took the film to be developed at local drugstores (or mailed it directly to Kodak), where it was turned into prints with Kodak chemicals and Kodak paper.

Filmless photography? You might as well call it *profitless* photography.

"Generally, the reaction to [digital photography] was that it was too far out there for serious consideration," said Sasson. "Not only was the camera not using film, but there was nothing in there that used any of the existing photo chain, the photo infrastructure that existed . . . throughout the world at the time. There was no photofinishing, there was no delivery—none of the things that we were used to thinking about. This was *really* far out there, and they couldn't grasp all of this."[9]

The rest of the story is now famous. By the time Kodak fully embraced digital photography, powerful competitors had already staked their claims in the arena. Kodak was never able to dominate the digital market as it had once dominated the film market.

Sasson's failure was not one of imagination, discovery, or experimentation, but a failure to overcome the objections of key stakeholders who insisted on clinging to a lucrative, but dying, business model. Although Sasson and his team did their best to battle the fierce headwinds, their navigation efforts proved futile in the end.

In 2009, Sasson was awarded the National Medal of Technology and Innovation by President Obama. Three years later, Eastman Kodak filed for bankruptcy.

WHAT IS NAVIGATION?

Where innovation is concerned, navigation is about dealing with the external environment and adjusting to the forces that can make or break your solution.

During the innovation journey, you may encounter people with motives to derail your mission, even after you've produced a working solution. The Kodak example proves that it takes more than a breakthrough prototype to succeed. Having produced a viable solution, you must now transform it into a practical product, service, or business. This is where navigation comes into play. Without buy-in from key decision-makers, your solution (often developed underground with input from trusted and/or supportive critics) may not survive first contact with the organization, investors, or industry partners.

Navigation is about bringing your solution to market—or to the beneficiaries—and scaling up. This requires you to build momentum and maneuver past dangerous obstacles. You must also meet the twin challenges of (a) keeping your solution alive while (b) ensuring that it's not adulterated to the point that it's no longer disruptive. You want your solution to remain original, despite all the forces trying to derail it or make it conform.

To accomplish all these tasks, you need to think differently. Methodologies like the business model canvas and stakeholder analysis are key preparatory instruments, but navigation also means reacting to unforeseen threats and opportunities as they arise.

PITFALLS

Whether working independently or within an organization, innovators often delude themselves in two ways.

First, like Steven Sasson, they overestimate the ability of their breakthrough solution to speak for itself and succeed on its own merits. They fall prey to the "if you build it, they will come" fallacy. Like the inventors of the Sony Reader and the Segway, they convince themselves that the demonstrable superiority of their offerings will prove irresistible to potential users, partners, and other stakeholders.

Such overconfidence is an occupational hazard for many entrepreneurs and intrapreneurs. After their initial success in championing their idea and winning over skeptics, they become good at convincing others. Thanks to this, some original thinkers succumb to hubris toward the end of the innovation journey. They are so accustomed to overcoming objections and converting resisters while ignoring intransigent naysayers (Laurence Kemball-Cook of Pavegen heard the word *impossible* from 150 venture capitalists) that they no longer heed warnings or suffer criticism. They take up residence in a reality-distortion field.

Second, they underestimate the potential hostility of the environment. The Segway team failed to anticipate regulatory pressures in many states and countries. The Segway's revolutionary design didn't fit existing vehicle categories, so it was often banned from sidewalks or roads. Similarly, Google Glass, which allowed wearers to surreptitiously film anything in their view, ran afoul of bans in establishments such as casinos and bars. As privacy concerns grew, the device also suffered from media references to its wearers as "Glassholes." In January 2015, barely nine months after it went on sale to the general public, Google pulled the plug on the product.

Revolutionary products and ideas can easily be hit by unforeseen or unforeseeable external forces. Such reverses are to be expected. But for innovators, what often comes as the biggest shock is *internal* opposition from the very organization that stands to benefit most from the breakthrough—as in the case of digital photography at Kodak. The detractors didn't kill the idea outright, but they sat on it for so long that they lost the advantage they should have had as inventors of the technology.

This is not unusual. Great solutions get quashed all the time, often because they don't fit the prevailing business lines or business model. Xerox's Palo Alto Research Company (PARC) is a veritable shrine to missed opportunities. It has a spectacular history of breakthrough inventions that were later seized by others, including the first true PC (the Xerox Alto), Ethernet networking, the graphical user interface, icons, bitmapping, scalable type, the computer mouse, and the world's first laser printer.[10]

Another risk is that your potentially great solution for the few will be slowly transformed into a mediocre solution for the many, as happened to Google Glass. Google reintroduced the device as "Glass" in July 2017. It's now positioned as a device for professionals, from doctors to automobile assemblers, that allows them to work hands-free.

As a result of these two delusions—overestimating the value of the solution and underestimating the need for buy-in—innovators often misjudge the effort required to mobilize supporters and overcome obstacles. They also fail to appreciate the need for originality in bridging that gap.

Adopting the ALIEN thinker perspective helps you redress the balance—to think simultaneously in terms of managing risk and gaining traction.

ALIEN THINKER NAVIGATION

ALIEN thinkers curb overconfidence by taking nothing for granted. They try to identify the forces, internal and external, that could potentially help or harm them. Adopting an ALIEN perspective alerts you to two key aspects of disruptive innovation that are familiar to practitioners but largely absent from the literature: (1) avoiding getting shot down, and (2) acquiring additional lift from the ecosystem.

ALIEN thinking sharpens your reading of a potentially hostile environment, including your organization's own immune system. It helps you identify both the likely backers and the likely blockers of your project. It draws your attention to unconventional pathways and sources of energy. It also equips you with techniques to anticipate and manage those forces.

To fully appreciate the complexity of navigation, you must split it up into its two constituent components: the need to *survive* and the need to *thrive*.

HOW TO SURVIVE

To navigate safely, start by reconnoitering the terrain, especially potential hazards, and then develop strategies to manage the hazards. Many breakthrough solutions flounder because the originators underinvested in anticipating the risks or negative forces. They overestimated the strength of their solution, and as a result, they collided with predictable obstacles. Survival is about anticipating the forces that can break your solution in order to avoid them. It also requires that you resist or neutralize less predictable threats.

Mapping Frictions

When you come forward with disruptive ideas, you *will* encounter friction. ALIEN thinkers know they must understand the rules (and the reasons for them) before breaking those rules. Innovators often underinvest in mapping the terrain and plotting a safe course. (The most direct route is not necessarily the best.) Too often, they charge into the thick of battle without knowing enough about their opponents' strengths. In this respect, the myth of the "crazy ones"—the mavericks who succeed by ignoring the rules and conventions—is both misleading and unhelpful. Many great ideas are undone by the failure of their originators to take full measure of the environment before leaping into action.

In the case of Sasson at Kodak, it's worth noting that he did foresee a key source of resistance, which centered on the camera's storage capacity. So he deliberately limited it to thirty images to match standard film rolls of twenty-four to thirty-six exposures. A familiar number. A comforting number.

Later, when producing the first experimental digital camera, his chief designer reminded him, "It's got to look like a camera; it's got to act like a camera and we're not going to pretend it's not a camera."[11] Half a dozen of these cameras were hand-built. They were never manufactured or advertised, "but we used them internally to see if anyone would be interested in marketing these things."[12] The intellectual property in that model—the D-5000 (or ECAM)—is licensed by every camera manufacturer in the world today. It is widely regarded as the precursor to most modern digital single-lens reflex cameras.

Unfortunately, these efforts were undermined by the way Sasson presented the new technology in meetings. Looking back, he conceded that dubbing the project "filmless photography" was a serious hindrance to gaining traction.[13] "The answer we got back from the channels was, 'Oh sure we would [market it], but if it comes at the expense of one film camera we won't do it.'"[14] Having resolved all the technical problems and laid the foundations for the future of photography, the company delayed in committing to the technology until others had closed the gap. And when Kodak finally *did* produce its first consumer camera, in 1994, it did so under the Apple brand. Apple marketed it; Kodak designed and built it. The example of Kodak dramatically illustrates the insidious power of the corporate immune system to resist game-changing solutions.

Resistance can flare up whenever you propose a solution that violates industry norms. Take the case of James Dyson. His ingenious bagless vacuum cleaner was turned down by all the major home-appliance manufacturers. Looking back, Dyson acknowledged his naivete: "I had visions of a vacuum revolution. Reality was something different. . . . These vacuum makers had built a razor-and-blade business model reliant on the profits from bags and filters. No one would license my idea . . . not because it was a bad one, but because it was bad for business."[15] It took Dyson several years to overcome these rebuttals, to fight legal battles on both sides of the Atlantic to protect his patents, and finally to raise the funds to start his own manufacturing facility.

Before presenting your solution, ask yourself, How will my solution clash with current arrangements—with the established order of this social system, organization, sector, or profession? Where are the points of friction?

A new solution can cause friction at different levels: internal (from the organization) as well as external (from the environment). It can, like Sasson's digital camera, challenge the structure, power hierarchy, technology commitments, and competencies of the organization—even if (as in Sasson's case) it is aligned with the purpose, values, and heritage of the company. Any breakthrough that threatens the status, job security, and career prospects of individuals is bound to trigger resistance, so you need to assess whose toes you might be stepping on.

The solution could also create disturbances within the ecosystem. It might alter relations between industry partners (like the Sony Reader), attract regulatory scrutiny (like Segway), or require changes in consumers' habits (like Google Glass).

Mapping sources of friction alerts you to likely obstacles. Always consider how your solution might upset established thinking, relationships, or practices—within the organization or within the ecosystem—in political, regulatory, economic, structural, social, psychological, technological, or functional terms.

Disruptions to these areas will mobilize people against you (once they detect you in the system), but understanding the underlying forces will help you predict which stakeholders are likely to resist. It will also explain *why* they may resist and help you forecast the intensity of the resistance, as well as the form, overt or covert, the resistance may take. This intel is critical to preparing your defenses.

Opposition can express itself in many ways, most actively through lawsuits, public criticism, or withdrawal of cooperation. But passive opposition, such as skepticism about your solution or false encouragement, can be equally deadly in sapping your energy or credibility.

When engaging with others, you need to be conscious that the same characteristics that make your solution distinctive also make it divisive. To lessen the frictions, you must find ways to frame the solution in a way that does not—like the words *filmless* or *bagless*—trigger allergic reactions. In the words of a former Lego innovation director, you need to be a "diplomatic rebel."[16] If you want to be disruptive, you have to build bridges. You need to think of ways to adapt your message and your actions so you can maneuver past the potential road-blockers.

Neutralizing Threats

Being aware of likely friction points will help you circumvent them. However, you may also have to take evasive action. You may have to maneuver to avoid being seen or avoid being hit. In some cases, though, you may have to maneuver in a way that lets you absorb hits with the least amount of damage.

One way to avoid taking hits is to keep the enemy from firing at you in the first place. To do this, use camouflage to evade detection and the destructive judgment of others. Make yourself small, and minimize the differences with your surroundings. This is a particularly useful tactic at the start of an innovation journey—before you have something to show.

One reason why Sasson managed to produce a working model was because he worked underground for so long. Initially, the project had no real budget and no allocated space, so he cleared out a lab at the end of a long corridor. The project escaped the notice of top management because Sasson didn't ask for money or support. Instead, he scavenged materials from the used-parts bin and stayed out of sight.

A recurrent theme among ALIEN thinkers working in organizations is the need to stay in "submarine mode"—below the radar, off the grid—so as not to activate the corporate immune system (disapproving bosses or colleagues) when the project is still weak.

Of course, you can't remain submerged forever. Once you surface, however, you can minimize the disruptive potential of your project by stressing its links with the past. You can demonstrate that, although the solution is different, it's not *radically* different. This approach makes it harder to brand you as a rebel. Annalisa Gigante, head of innovation at LafargeHolcim, the world's largest cement company, cites this as her biggest lesson (from painful experience) in pushing innovative projects: "The closer you can propose an idea that really matches with the goals of the company, the easier it is. . . . So it's really understanding what the core of the company is. If it's sustainability and if it's an idea that increases sustainability, then you will already have that momentum with you. . . . It's about packaging and finding output metrics that will link to what the rest of the team is measured on."[17]

Within your organization, you need to present a disruptive innovation in a way that is responsive to the collective DNA. According to Jean-Paul Bailly, who oversaw a dramatic transformation of the French national mail service (La Poste) from 2002 to 2013, "You have to demonstrate that the change can help you remain true to your identity."[18]

In 2010, even as the state-owned mail service began to privatize, Bailly continued to emphasize public service values and public trust. Those core assets underpinned an array of new activities, including e-commerce, banking, and mobile telephony. Building on public confidence in mail carriers, the organization added services for the elderly—an innovative response to the country's aging population and the declining number of letters sent by post. Customers can now commission the local mail carrier to drop in on aging relatives, and postal workers are trained in how to respond to the situations they might encounter. Thanks to such measures, La Poste's transformation was completed without any layoffs, and the company's revenues have continually grown.

The same principle—linking your radical project with familiar and accepted concepts—applies externally, too. For example, with his drone-based air cargo company, Jonathan Ledgard was careful to soften his message when he proposed using drones to address Africa's emergency-supply challenges. Anticipating a risk of rejection because so many people associated drones with missile attacks, Ledgard called his drones "flying donkeys."[19] This unexpected euphemism, borrowed from a Kenyan farmer, captured the idea perfectly, making it concrete, unthreatening, and sticky.

Similarly, when creating a credit score for the formerly incarcerated, Teresa Hodge designed it to range from 300 to 850—just like the standard FICO credit score that decision-makers used to determine whether you can secure a loan, buy a car, get a job, or rent an apartment. The way you frame your solution may be vital to its survival.

Resisting Attacks

You can't anticipate and circumvent every threat, so you must also adapt. Once you become visible, you may need to be agile and sometimes switch tactics. These defenses may not keep you from being shot, but they may prevent you from being *shot down*.

Take the case of Owlet, a health technology company founded in 2013. The start-up devised a wireless wristband to monitor hospital patients' vital signs. The development team thought it had a winner because neither patients nor nurses liked the wired devices currently in use in hospitals. But the team ran into unexpected opposition from hospital administrators once they introduced their product. Among nurses, the most complained-about issue was tangles of wires, but because this wasn't a pain point for administrators, they refused to pay for the new wristbands. In focusing on users, Owlet had neglected the buyers.

They quickly redirected their efforts and adapted the same technology for use in the home instead of a hospital setting. They developed a "smart sock" that tracks a sleeping infant's pulse and breathing and, when warranted, sends an alarm to the caregiver's smartphone. This time, crucially, the buyers (parents) also received the main benefit (peace of mind).

But once again, the team at Owlet ran into opposition—this time from the US Food and Drug Administration (FDA). The issue was the alarm. As one of the founders explained, "Alarm plus pulse oximetry means you need FDA clearance. . . . Because it sounds an alarm, that means that we are giving advice to parents—we're saying, 'Hey, there's something going on here.'"[20]

This time, the Owlet team quickly tested and found a consumer appetite for a non-alarm version of the product that only tracked pulse and respiration. This kept the company in business through the lengthy process of FDA approval, until it was able to release its upgraded flagship product with an alarm.

Owlet is a strong reminder that even a breakthrough solution exists within a system of interdependencies, not in isolation. Twice, Owlet was blindsided by a value-critical stakeholder it had overlooked. But it was able to deflect the hit on both occasions. Fast reflexes and mental agility are key to surviving unexpected blows and bouncing back without losing too much momentum.

Unorthodox defensive moves are another option for ALIEN thinkers. These can be used not only to deflect hits, but to actually leverage them.

Take the case of TerraCycle, a for-profit recycling business.

Canadian student Tom Szaky dropped out of Princeton in 2002 to produce and distribute an organic fertilizer branded as Worm Poop, which later gave rise to the business TerraCycle. He collected food waste from the university and fed it to worms, which turned it into rich fertilizer that could be liquefied and packaged in used soda bottles.

TerraCycle's big break came when it won first prize at the Carrot Capital Business Plan Challenge, which came with a $1 million investment in the business. The only catch: the investors wanted to orient the business toward a wider line of organic fertilizers and tone down the environmental message that had become synonymous with the start-up. It was a serious blow to the company's distinctiveness. With only $500 in the bank, TerraCycle desperately needed the investment. But Szaky turned it down. Instead, he leveraged the interest shown by Carrot Capital to attract alternative funding and gain attention from the biggest retailers, rather than continuing to develop the brand through small stores and nurseries, as his original potential investors advised.

Within a year, Szaky persuaded Home Depot Canada to start stocking his product online. It was not a big order, but it opened more doors. Walmart Canada and other chains followed suit. Of course, once TerraCycle started to compete for shelf space, the competition took notice. Scotts Miracle-Gro, the behemoth of plant food, filed a lawsuit claiming that TerraCycle's packaging resembled Miracle-Gro's too closely and was confusing consumers. Though far-fetched, given that TerraCycle's packaging was used soda bottles, not sleek containers, Scotts's claim had to be answered.

TerraCycle risked running out of money. Even if the lawsuit didn't bankrupt them, having to respond to the lawsuit risked distracting them from the work of meeting shipping dates, which could undermine the credibility they had started to establish with the large retail chains.

So TerraCycle decided to use Scotts's dominance to their advantage. "We started a Web site, suedbyscotts.com, that was intended to get our side of the story out," said Szaky.[21] Beyond stating the details of the complaint—and cheekily asking people if they could tell the difference between the two products—the narrative painted a David vs. Goliath struggle that was tailor-made for the news media: "We were a 'Small, Eco-Friendly, Organic Company Started by Students.' Our bottles were recycled, collected by 'children in communities across the land.'"[22]

The net result? "Ninety days later, 150 articles had been written, from the cover of the *Wall Street Journal* to *BBC World News*, and Miracle-Gro ended up settling on very favorable terms."[23] The internet-based defense did not just shame Scotts into withdrawing its lawsuit; it also brought TerraCycle an unprecedented level of publicity. They came out of the lawsuit strengthened. It put them on the map as the "best-known organic fertilizer product in the country."[24]

HOW TO THRIVE

When pushing your disruptive solution into the world, your first priority must be survival and risk management. But in order to *thrive*, you must find ways to grow stronger and more resilient—to keep developing and improving.

If you stay in survival mode, the danger is that you'll lose time and make too many compromises. Alongside the risk of getting shot is the risk of losing your edge or originality. Your efforts to blend in might make you bland. To thrive, you need to identify positive forces that can help your breakthrough idea gain traction, and you need to find original ways of harnessing those forces to ensure that everyone benefits.

As TerraCycle's growth accelerated, Szaky realized that they had stumbled on a model that could also be applied to trash other than food waste. Garbage was a commodity with negative value—in other words, people sometimes paid others to dispose of it. And fertilizer represented only a tiny fraction of the market they could exploit. There were many other untapped types of waste.

So he decided to shift the focus of the company from the product— plant food—to the commodity—garbage. His thinking was, "Instead of making the product the hero, let's make the garbage the hero and figure out, 'Okay, whatever the type of garbage is, how do we collect it? How do we process it? And how do we make a business model work?'"[25]

The novelty of his approach lay not in targeting commonly recycled items, such as glass, metal, and plastic, for which there was a straightforward business case, but in targeting waste products that others neglected. Szaky saw opportunities everywhere.

"We started running [collection] programs for brands like Honest Tea, CLIF BAR and Stonyfield Farm," he recalled. "Within a year we were working with Kraft Foods brands like Capri Sun and Nabisco, with Frito-Lay and with Mars. It was clear our new model was ripe with opportunity."[26]

There are hardly any products that cannot be recycled. The problem is that it usually costs more to collect and process waste materials than they're worth. TerraCycle could expand their existing collection programs to include other waste, but the big challenge was finding partners willing to fund and support the effort needed to recycle materials that were not yet recyclable.

TerraCycle's initial approach didn't have much impact. They asked companies to participate because it was the right thing to do, but appeals to altruism got them only so far. Budgets were limited, and the programs were time bound.[27] Soon, they realized that profits were a better motivator than guilt: "Our job is not just to go in and say, 'You should pay for this, because you have some theoretical responsibility.' Much more, it's 'You should pay for this, because this is how you're going to drive value.'"[28] In other words, Szaky decided to use an ROI-based argument rather than one stressing the benefits to society or the planet. "The challenge is that most organizations that do sustainability work phrase it as 'You should do it because it's the right thing to do.' While it *is* the right thing to do, that framing is what makes it usually not scale."[29]

Since changing their approach, TerraCycle's growth has been spectacular—in fact, they've given birth to a whole new industry. TerraCycle now operates in twenty-six countries around the world with a licensing model. Beyond North America, it has a presence in China, Japan, Western Europe, and Latin America. It has been featured three times in *Inc.* magazine's list of the top one thousand fastest-growing privately owned companies, and it has twice had its own reality TV series—the shows *Garbage Moguls* on the National Geographic Channel and *Human Resources* on PivotTV—which provide what Szaky calls "negative-cost marketing." Why pay for advertising when you can get paid to be the content?

Szaky has all the hallmarks of an ALIEN thinker, and his efforts illustrate the three key aspects of navigation that you need to thrive: mapping overlooked opportunities; finding unconventional means of leveraging those opportunities; and seizing on unexpected opportunities that present themselves.

Mapping Overlooked Opportunities

You need to identify opportunities to configure an original business model and achieve scale with unconventional partners.

For TerraCycle, the overlooked opportunities were plentiful because they chose to focus on products that were not economically profitable to recycle. Instead of picking the very best garbage, they targeted the toughest types of waste, including cigarette butts, dirty diapers, and chewing gum. They focused on what no one else was doing.[30] To exploit these opportunities, however, Szaky had to seek partners beyond TerraCycle's existing network of schools and environmental groups (which did the collecting) and retail outlets (which sold the products). He had to find stakeholders willing to fund the recycling, and that meant approaching large companies as sponsors.

So Szaky reached out to producers of some of the world's most littered items, especially in consumer goods. He proudly noted in 2018, "Today at TerraCycle, we work with every major tobacco company without exception. Big oil, big pharma, big food, you name it. . . . It's also about scale. If I can go in and affect . . . any of these large organizations, the change is much bigger."[31] Volumes increase and the total net effect is amplified.

When the obvious partners turn you down, ALIEN thinking helps you identify other candidates. For example, Bertrand Piccard found that the aviation industry had zero interest in supporting his efforts to build a solar plane. They told him it was impossible to produce enough energy to keep the plane aloft at night. So he turned to yacht builders instead. "If people believe something is impossible, you have to find people who don't *know* it's impossible," said Piccard. "And that was a shipyard who told us, 'We can make your plane as light as you want because we know how to use carbon fiber.'"[32]

The new partner enabled a paradigm shift from the challenge of producing more energy to that of consuming less energy. Ultimately, this led Piccard and his team to design "a plane with the wingspan of a jumbo jet but the weight of a family car."[33] Novel collaborations are often critical to delivering breakthrough solutions.

Another great example is Vestergaard Frandsen (VF), the Swiss-based disease-control company. They devised an ingenious line of water filters that sold well to hikers and campers but were far too costly for the rural African and Indian communities that needed them most. Because the filters reduce air pollution by eliminating the need to purify water over open fires, VF hatched the idea of distributing the devices for free and using funding from carbon offsets to generate revenues.

To unlock this novel source of funding, the company had to satisfy independent auditors that hundreds of thousands of filters were actually being used. So VF looked to an open-source data collection platform developed at the University of Washington, and worked with the platform's developers to create a smartphone app that would allow field representatives to photograph recipients of the filters and record their homes' GPS coordinates. Now, each recipient was reachable for follow-up and auditing purposes, making the solution scalable *and* sustainable.

Both VF and TerraCycle blur the boundaries between for-profit and nonprofit organizations. To identify unconventional solutions and partnerships, you can learn a lot from the ingenuity of frugal and social innovators. Deprived of resources, they must find alternative ways to engage with the private sector, governments, NGOs, and banks to scale the benefits of their innovations to large populations. For example, health-care firms in Africa are piggybacking on Coca-Cola's cold chain to preserve and deliver critical medicines to remote villages.[34]

Leveraging Opportunities

Of course, identifying promising partners is one thing; convincing them to be energetic contributors is another. As an initiator of breakthrough solutions, you need to find creative ways of aligning the interests of partners, as well as responding to unexpected opportunities that may emerge from the ecosystem.

To align its interests with prospective stakeholders, TerraCycle had to craft win-win propositions. Once it understood the need to "embrace the ROI conversation"[35] to inspire partners to get involved, TerraCycle was careful to adapt its messages to different stakeholders—and to build bridges by talking their language.

TerraCycle identified five core stakeholder groups who could become funding sources:

1. Consumer product brands (e.g., Colgate, L'Oréal) that fund the recycling of their waste
2. Retailers like Target and Office Depot, which collect items such as car seats and old binders, respectively
3. Municipalities, such as New Orleans and Vancouver, that operate cigarette or chewing gum recycling platforms
4. Manufacturing facilities interested in moving to zero waste
5. Small offices and individuals who can purchase zero waste boxes to recycle mixed waste

With each stakeholder group, the pitch had to be different. TerraCycle had to adjust the message to emphasize how recycling would benefit that *particular* partner. "So, if it's a retailer, how will it help them sell more stuff; if it's a consumer product company, how will it help them beat their competition; if it's a city, how will it help them get more tourism or tax revenue?" said Szaky. "If you frame sustainability in that mode, you will get to the finish line much quicker and may attract bigger budgets and a longer-term commitment from the folks that you're speaking to."[36] In other words, you need to figure out how you can benefit the prospective partner in a traditional way. "It's not just saying, 'This person or that company should pay'; it's getting them to *want* to pay because it serves their core business. And figuring that out is the entire key challenge to our business."[37]

Don't try to outmaneuver potential partners; inspire them to collaborate. Among the coalition of partners that Piccard drew to the Solar Impulse project, he was able to leverage a shared interest in what would be learned during the process, regardless of the outcome. "Several companies told me, 'We don't know if it's possible, but we want to try with you, because it will stimulate innovation internally . . . and motivate our teams to be more innovative and to get into this process of disruption.'"[38]

Piccard took a similar approach with the sponsors of his project: "When you stop *convincing* and you start *motivating*, you make the alliance with the part of the [funder] that would like to say yes. And you stimulate this part; it's not a fight anymore—it's an alliance. And this is the moment where you understand that it should not be a sponsor that is paying for your dream; it should be a *partner* who is really interested to be a part of the project."[39]

The challenge is finding creative ways to bridge goals in order to achieve traction. A good example is the Mexican-born theme park chain KidZania, an indoor "city" where children can role-play adult jobs. When its founders ran out of funding to launch the concept, they decided to approach corporate sponsors, including DHL, Nestlé, and HSBC. Rather than just asking for financial support, they also leveraged their sponsors' specific industry knowledge to help design the activities and infrastructure for a more realistic experience.[40] The concept was an instant hit. Within two years, the park was attracting twice the number of visitors initially forecast.[41] And it has since become the world's fastest-growing group in experiential learning for children. It now has twenty-four parks in nineteen countries across five continents, with another twelve under construction.[42]

Cofounder and CEO Xavier López Ancona, a former GE Capital VP, said, "The business model is based on a win-win scheme: for our customers who are children; for their parents, who we help by reinforcing values; similarly for schools; for the malls for whom we are a differentiator; for the brands that are shared, it is a way to approach customers; for the work team; and, of course, it is a good deal for investors. . . . Everybody wins."[43] KidZania managed to turn its sponsors into partners *and* cocreators.

Chris Sheldrick, cofounder of What3words, also talked about having an open conversation with potential partners and investors: "Not overhyping things too much and just being really straightforward—'This is what we're doing, we're really committed to it, we believe we can do it, do you want to come with us?' You have to get away from a mentality of us-and-them."[44]

Labels like *sponsor*, *supplier*, and *retailer* tend to limit what you expect from partners and what you think you can offer them in return. They encourage a mindset of extraction rather than synergy.

Seizing Unexpected Opportunities

In addition to connecting with unconventional partners, you need to exploit unexpected opportunities. Navigation, in contrast with implementation, sometimes requires improvisation and daring, not just execution. You may even have to relinquish full control for the solution to flourish.

After the company's success with two KidZania parks in Mexico, the cofounders became interested in international expansion. However, they disagreed on whether to start in the neighboring US. As a consequence, they split up. Luis Javier Laresgoiti sold his share to

López and moved to Florida to launch Wannado City, a theme park similar to KidZania but three times the size. López took more seriously the advice of several experts who recommended that he delay entering the US market, among them Howard Schultz of Starbucks. "First, grow with your brand in Mexico, then try it in another market . . . and in another," warned Schultz. "Doing it in the U.S. too fast can break you."[45] As López developed a franchise plan to expand abroad, a Japanese entrepreneur took an interest in the project. Within three months, the company signed a contract to build the first franchise in Tokyo.

It was an unconventional move for a Mexican company. Those not expanding into the US typically headed south into Central America, or possibly overseas into Spain. KidZania was only the second Mexican group to invest in Japan (after cement giant Cemex). But López realized that Japan offered access to a huge market, and the country was also seen as a trendsetter by other Asian nations: "We have to go where the cities are, and 60 percent of the population is in Asia."[46] It proved to be a shrewd move. First-year attendance at the Tokyo franchise was more than twice that of the Mexican parks.[47]

While continuing to manage the two (and later three) parks in the Mexican market, López created a separate unit to manage the franchises and pursued rapid growth across major Asian cities. Additional franchises were opened in Jakarta, Seoul, Kuala Lumpur, Bangkok, Mumbai, Kuwait, Cairo, Istanbul, and Jeddah, as well as Lisbon and London. Working with local operators as well as international partners, KidZania was clearly a business concept that could span borders and cultures.

For the franchising model to work, López underlined the importance of "documenting the business, doing the manual, knowing the industry, looking for the appropriate partners and territories."[48] López himself focused on identifying appropriate partners and territories for the concept. And he insisted on carefully selecting partners, both prominent local companies and global ones, who shared KidZania's joint commitment to learning and fun. Parks were always located inside important malls, where people welcomed protection from the rain or heat, and in cities of over five million inhabitants with a concentration of corporate headquarters and good transport links to the venue. They also targeted cities with few rival offerings and, therefore, plenty of demand to satisfy. This enabled them to build their competencies as they continued their expansion around the world.

Today, the company is expanding into the US market. The Wannado City venture, established in the US by López's former partner, closed in 2011 because of financial problems and lack of business. By contrast, López has carved out a new niche in the entertainment industry—the so-called edutainment sector. He understood that KidZania was not meant to be a destination venue like Disney's parks but rather a place close to the community that children might visit repeatedly.

Looking back on Howard Schultz's early recommendation about building successful franchises outside Mexico before targeting the US, López smiled. "It was quite good advice."[49]

The KidZania case offers two key lessons for ALIEN thinker navigation: (1) be prepared to improvise, and (2) be prepared to let go to grow. At KidZania, the improvisation involved seizing opportunities to go where the energy was and where they had an eager partner, starting in Japan. Letting go required López to opt for a franchise model rather than trying to retain full operational control of the international parks.

At TerraCycle, Szaky sought to harness organic forces and seize natural opportunities instead of rowing against the tide. Referring to his expansion strategy, he said, "It's opportunistic. It's where there's interest. So, in China, we're targeting oral care recycling and cosmetic recycling, but it could be anything. It's truly where there's opportunity and where there's interest to fund solutions."[50]

When it comes to relinquishing control, he noted that the "TerraCycle business model is very unique in that collecting and repurposing material like ours had never been addressed before. However, we've always remained flexible with our business model, goals, mission, and how we create products. We've moved from a manufacturing model to a licensing model, which has helped make TerraCycle profitable. I think that that flexibility is key for any young business to succeed. They can't be afraid to change and adjust."[51]

Looking back on TerraCycle's early history, Szaky stressed the navigational effort required: "It would have been impossible to predict or plan how to develop TerraCycle to the place where it stands today. The trick was to be ever vigilant in seeking opportunities, and to be ready to jump on them if they felt right inside and consistent with our core mission, even before they could be well thought out."[52]

DIGITAL NAVIGATION

Digital instruments and channels can enhance the navigational skills of ALIEN thinkers, as they did for Narayana Peesapaty and his edible cutlery when his story went viral (see Chapter 2). In addition, they enable ALIEN thinkers to more quickly interpret the environment and reduce their chances of being blindsided.

Digital technology can also provide opportunities to develop fresh business models or connect with unconventional allies, enabling a breakthrough solution to attract attention and thrive. Recall how VF used digital support to gain access to an original source of funding— carbon credits—for its water filters. The company issued smartphones to field representatives in Kenya and cocreated an app that allowed them to take photos of the installation and record the name of the beneficiary and the number of people in the household, as well as the GPS coordinates. This data allowed independent auditors to verify the information submitted, so the company could receive the carbon credits.

Clearly, digital technologies offer a huge boost in terms of scaling, as well as a means to continually improve the quality and marketing of your solution. For example, an innovation launched at one of KidZania's Mexican operations is digital passports (for those who want them), complete with photo and date of birth. Each passport is stamped on completion of an activity. The passports provide López and his team with detailed feedback on children's interests and tastes—which activities work best with which segments of the audience—and help the company improve its marketing to the adults responsible for the children.[53]

Crisis Text Line, a nonprofit organization providing counseling to vulnerable teens through text messages and social media, launched a subsidiary site called Crisis Trends that makes anonymized and aggregated data freely available to researchers, public health agencies, and the general public. Rather than hoard the data or sell it, it has chosen to use the data to make the world a safer place by attracting world-class machine learning researchers and experts in data applications to address the challenge of improving population mental health. Crisis Text Line didn't have the in-house skills to navigate the process from data to insights to action, so it invited others to collaborate.

An Antidote to Overconfidence

As we mentioned earlier, ALIEN thinkers face two common obstacles. First, they tend to overestimate the power of their breakthrough ideas. Second, they tend to underestimate the resistance they will face. Before ideas can survive and thrive, these obstacles need to be surmounted. Digital tools and technologies can help overcome both obstacles.

Overconfidence often stems from a lack of external feedback. The more we immerse ourselves in our own ideas, the more convinced we become of their value. Digital tools can serve as an antidote to overconfidence by providing faster feedback, from a broader swath of potential customers.

Ideafoster, based in Barcelona, is a consultancy that provides quick testing of breakthrough ideas with real people via Facebook. They take an idea, such as a product prototype or new business concept, and create multiple Facebook ads that pitch the idea in slightly different ways. The ads are tested with different stakeholder profiles using Facebook analytics. For example, an ad might be designed to appeal to, say, professional women, ages thirty to thirty-five, married, with children, and living in London. The ads are then tested with thousands of Facebook users who fit that profile. When users click on an ad, they are directed to a landing page where the new concept is presented. Feedback is quickly received and analyzed, often within twenty-four hours. Such feedback can be used not only to improve the offering, but also to support the "sale" of the solution within an organization.

Ideafoster and companies like it provide fast, scalable feedback mechanisms for breakthrough ideas. However, they are not the only way to collect feedback online. Question-and-answer sites such as Quora, Ask, and Answers.com can also provide answers to specific questions. The benefit of these sites is that real people will respond with specific, detailed, and often brutally honest feedback, mostly for free. These questions may focus on the breakthrough idea itself—such as its strengths and weaknesses—or on how it's likely to be received by organizational stakeholders.

For example, an idea about a new way to process travel receipts could be targeted to accounting professionals and controllers with a question like, "If someone suggested this idea to you, how would you react?" By soliciting answers from anonymous critics—from people who live outside your circle of friends and colleagues—you are more likely to receive blunt and honest feedback.

Online feedback can bring previously invisible issues into stark relief. For example, Uber struggled in Southeast Asia in part because it failed to offer a cash payment option. The fact that Uber was cashless was considered a key benefit in Western countries, but it was a liability in areas where credit cards and other non-cash options are not as widespread. Local competitors such as Grab and Gojek were quick to address this need.[54] Had Uber tested its concept online in the markets of Southeast Asia, it might have been quicker to appreciate the importance of offering a cash payment option, even though this deviated from its global standard.

Fast online feedback from a variety of external stakeholders offers a strong reality check for ALIEN thinkers who might otherwise succumb to overconfidence about their offerings.

Quelling Resistance

To overcome resistance to your ideas, consider implementing one (or even all) of the following four digital strategies.

Exploit the Digital Hype

There is massive global interest in digital disruption and transformation. It's difficult to browse a news website, read a trade magazine, or visit an industry conference that doesn't include a significant focus on digital. Indeed, there's a high level of interest in digital topics, and increasing concern about the impact of digital disruption, across most sectors of the economy. Research tells us that creating a sense of urgency is an important step in overcoming resistance to any change initiative.[55]

As an ALIEN thinker, you can take advantage of the heightened attention surrounding digital to generate more interest for your idea. Adding a digital veneer to your idea can help it gain the necessary credibility to be taken seriously by an organization. This isn't to suggest that you should misrepresent an idea—only that you can exploit the strategic prioritization of certain issues (in this case, the heightened attention to digital within many organizations) to your advantage.

Viral Campaigns

Often, resistance comes in a very specific form: a request for validation. This need for proof presents a problem for the ALIEN thinker, as new ideas, by definition, are not mature enough to have much external validation. Fortunately, the online world can provide fast validation—which can be used to both reduce overconfidence and overcome resistance.

Viral campaigns can heighten the impact of new ideas. For example, Procter & Gamble struggled for years to update the image of its Old Spice brand, but nothing could alter the product's image as "my grandfather's aftershave."[56] A number of attempts to freshen up the brand's image failed—until a 2010 campaign by the ad agency Wieden+Kennedy, which was dubbed "the man your man could smell like," took off. One key reason for the campaign's success was that it targeted women (who often *purchased* the product) rather than men (who used it). Ironically, the tables were turned on Procter & Gamble two years later when a new competitor, Dollar Shave Club, launched its own viral marketing campaign that targeted Gillette, a Procter & Gamble brand.[57]

Viral campaigns can effectively generate the proof that skeptical executives need to support a breakthrough solution.

Psychographics

Psychographics provides a new take on market segmentation. Rather than segmenting by demographics—age, gender, socioeconomic status, and so on—psychographics segment by personality profile. Thanks to this, your messages can be adapted to match the personality profile of the recipient. And digital tools offer a new way to measure personality profiles—one that doesn't rely on traditional methods linked to personality tests.

Psychographics can be a potent weapon against resistance. Steven Sasson struggled to sell the idea of the digital camera, in part, because the term *filmless photography* rankled many key decision-makers at Kodak. If only Sasson had had access to the insights from the personality profiles of those executives, he could have modified his pitch accordingly. Some executives might have been emboldened by the idea of filmless photography (though others would have still been threatened by it).

Analytics company PatientBond has developed a solution to increase the percentage of people who get a flu shot by leveraging the fact that personality profiles respond differently to different types of messaging. For example, "self-achievers" (24 percent of the patient sample) respond to messages that stress goals and achievements, such as, "Your flu shot is a critical part of keeping you in charge of your health and achieving your health goals." By contrast, another psychographic segment, "direction-takers" (13 percent of the sample), tend to respond more favorably to the message, "Please get this flu shot, as medical experts strongly recommend it for staying healthy."

If you assess the personality profiles of the people you need to persuade and then adjust the message, the likelihood of overcoming resistance dramatically increases.

Reverse Mentoring

In 2016, Vera Bradley, one of the world's largest luggage and hand-bag retailers, began a social media campaign directed toward female customers. Designed to celebrate being a woman, the campaign included messages such as, "Treat yourself to a bouquet of flowers" and "Needing, not having, five shades of red lipstick." Customers were encouraged to send messages celebrating all things feminine with the hashtag #itsgoodtobeagirl on Twitter, Facebook, Instagram, and Pinterest.

The approach was bold, creative, and fresh. And it failed miserably.[58]

Instead of feeling celebrated and empowered, women, particularly the young millennials whom the company coveted, regarded the campaign as condescending and old-fashioned. Vera Bradley's social media channels were bombarded by comments such as, "Your #itsgoodtobeagirl campaign is 50s housewife propaganda" and "Forget about gender pay gap, institutional sexism, and sexual harassment. It's OK. We have 5 shades of lipstick."

Sales dropped.

Almost anyone under the age of thirty could have told Vera Bradley that the campaign would fall flat. Anything that calls out differences between the sexes is seen by most twentysomethings as misogynistic, patronizing, and inappropriate. How did Vera Bradley not know this? The short answer is that they didn't ask the right people.

One promising approach to addressing this issue is reverse mentoring. Reverse mentoring is a simple concept: you take traditional mentoring and turn it upside down. Instead of a seasoned executive mentoring young talent, the young talent mentors the executive. Launching such a program may open the minds of some executives to your new ideas.

In a 2019 *Harvard Business Review* article, IMD professors Jennifer Jordan and Michael Sorell described the challenge New York–based bank BNY Mellon's Pershing faced in attracting and retaining new generations of digitally savvy employees.[59] Part of their problem was that the leaders of the bank had very little knowledge of (or interest in) digital tools and technologies. The bank decided to set up a reverse mentoring program to pair leadership team members with a selection of young employees. One consequence of the program was that the bank's CEO (along with his young mentor) launched a series of fireside chats to enhance his connection with employees. Another senior leader became very active on social media, transforming himself from a digital Luddite into a key contributor to the bank's image across different channels like LinkedIn and Twitter.

PUTTING ALIEN THINKING TO WORK

Rebooting Logitech

Founded in 1981, Swiss-born Logitech became widely known in the early 2000s as a developer of elegant computer accessories, such as mice, keyboards, speakers, webcams, headsets, and gaming devices—products that are highly prone to commoditization, making continual innovation a matter of survival.

Surrounding the personal computer with clever devices proved a winning strategy. Logitech went on a streak of thirty-nine quarters of double-digit growth (both revenues and profits). Then in late 2009, everything changed with the launch of the iPad. Tablets exposed Logitech's complete dependency on a slowing and soon-to-be declining PC business. After three years of failed initiatives to launch new growth engines, Logitech brought in an outsider, Bracken Darrell, to lead the push into new product categories.

He started by inviting everyone at Logitech to speak up, and he listened. Through those early discussions, it became clear that Logitech had no problem coming up with breakthrough concepts. But steering disruptive ideas through the organization remained a challenge. Even in a company committed to innovation, good ideas sometimes got bogged down or worn down by internal mechanisms.

We met with Darrell to discuss how he tried to identify those inhibiting factors and help protect fragile but promising ideas.[60]

The Pull of Gravity

The biggest killer of radical ideas, according to Darrell, is the pull of organizational gravity. "Any radical idea that comes into the orbit of an established organization gets pulled down." The heritage of the company, its distinctive capabilities, and a strong sense of "the way we do things here" make it difficult to take off in a new direction. At Logitech, the focus was traditionally on hardware and devices.

Darrell recalled the example of an entrepreneur who created an app and was recruited to develop new software and service offerings. Instead of creating a new experience, the newcomer succumbed to the dominant paradigm and produced a piece of hardware. "It just shows you how powerful gravity is," said Darrell. "It sucks everybody into it."

Another aspect of Logitech's high gravity was the friction between functions, businesses, and regions. It could be marketing refusing to leverage customer insights gathered by an engineer,[61] or resistance from subsidiaries in lead markets to successful products developed for "less sophisticated" markets. "There can be a kind of unspoken barrier that creates resistance," noted Delphine Donne-Crock, Logitech's general manager of creativity and productivity. "People think if it didn't come from their group, it's not as good."[62] The not-invented-here syndrome clearly applies internally as well as externally.

The blockers turn out to be not just the bosses and gatekeepers with control over strategic choices and funding, but also the colleagues whose buy-in cannot be taken for granted. Moreover, aspiring innovators are often ill-equipped to combat or find their way through the organizational maze. "Knowing who to talk to is already a huge hurdle," said Donne-Crock. "If someone comes to you with a great idea that is unconnected to your area, it's like, 'Okay. How does that fit with my priorities or objectives?'"

In other cases, good ideas are lost because disrupters focus on the wrong arguments in their pitch. "People are confused in what they try to say the first time," said Donne-Crock. "They try to either say too much about the technology itself—or too much about the potential revenue of it. But most of the time, people don't talk enough about the user experience. Sometimes the 'cool experience' the innovator describes does not fit with the users being targeted."

Driving Better Navigation

Over time, Darrell has responded to these challenges in three ways.

Emulating Start-Ups

To reduce the effects of organizational gravity, the weight of the product empires, and functional silos, Darrell restructured the company.

Convinced that a well-run start-up was the best unit of design and source of innovation, he started breaking down the organization into smaller entities. "When I came in, we had really two large business groups," he recalled. "Today, we have twenty-seven different people running small businesses." He transformed the structure from big blocks to an aggregation of small teams, each tightly focused on a targeted population. Darrell called this approach "scaling small," the aim being to help Logitech act more like a small company—no longer a battleship but a flotilla of small, fast vessels able to react quickly to opportunities and disruptions.

Darrell has also given his "start-ups" more autonomy, making it easier for disrupters to work with outside partners if they can't marshal internal support. He is a strong advocate of the right to look outside for help. What used to be an exception is now a legitimate choice—notably when the capacity or resources are not available in-house or when external partners are willing to fund the development in exchange for a cut of the revenues.

Instilling a Common Focus

Design at Logitech was traditionally outsourced to external firms. One of Darrell's first decisions, in 2013, was to reallocate a significant chunk of R&D funds to build an in-house design capability, now 150 employees strong. To lead it, he hired Alastair Curtis, Nokia's former chief designer, who helped transform cell phones from a tool into a cool accessory.

Introducing design as the primary driver of the business helped to reduce turf wars between the traditional powerhouses: engineering and marketing. The initial impact was felt in new product development, where Logitech started systematically including representatives from design, engineering, and marketing on each innovation team. As Donne-Crock explained: "They all need to think about what's the best experience we could deliver, and they can all challenge each other with their respective expertise."

To avoid innovation blockages based on misguided assumptions, Donne-Crock stressed the "people first" principle. "It's not just marketing that needs to know the consumer. Even an engineer who comes with a great idea needs to take the time to figure out why it matters for a consumer, and how he or she knows that it matters." Everybody needs to come with data.

Guiding Disrupters

For their ideas to gain traction, innovators must find the right sponsor and clearly articulate why consumers need their idea.

At Logitech, helping disrupters gain exposure to sponsors starts at the top. To make it easier for innovators to knock on the right door, Darrell goes out of his way to make himself accessible and encourages others in the top team to do likewise. "One of the things I'm good at is connecting with people and listening to their early ideas. And Alastair [Curtis] is the same way. The two of us are a little bit like sponges, constantly soaking up ideas. And then, maybe a fifth or a tenth of them get to the next stage. But at least they get heard."

Helping disrupters gain attention for their ideas is also a preoccupation for Donne-Crock. She has developed a set of essential questions to guide disrupters through their pitch: "Is it a new use case? Is it a new experience? Is it a new user? Is it a new business model? And what group do I fit in? Is it about gaming? Is it about video collaboration? Is it about work-based productivity? If you can't answer certain very specific questions, then you're going to be lost and turning around in circles for a while." Disrupters are advised to develop different pitches for audiences rather than relying on the same presentation for all.

Navigation as a Capability

Darrell's redesign of Logitech took a little while to gain traction, but by 2020 profit levels were five times higher than they were in his first year (2013), and the stock value had gone up sevenfold. The company successfully negotiated the transition from PC platform to cloud platform while also expanding its product portfolio into new areas.

Darrell used his ALIEN perspective to question the organizational barriers to divergent ideas. He realized that innovation efforts in new categories needed more autonomy and different performance metrics, but also more attention: "You've got to give them way more attention than they deserve based on their size, or maybe even more than they deserve based on their potential. Because you will learn more from them than they will from you."

In the process, Darrell built up Logitech's navigation capability to the point that it is now better able to steer divergent ideas to commercialization—and not just internal ideas but also ideas from outside. As he put it, "Start-ups frequently hit the wall like little insects, and we're there to scoop them up before they're gone, bring them into our company, bring them back to health, and turn them into something special."

KEY TAKEAWAYS

- Navigation is about dealing with the external environment and adjusting to the forces that can make or break your solution.
- Without buy-in from key decision-makers, your solution may not survive contact with the organization, investors, or industry partners. Steve Sasson's experience at Kodak proves that it takes more than a breakthrough prototype to succeed.

- When it comes to navigation, innovators often delude themselves in two ways: (1) they overestimate the ability of their breakthrough solution to speak for itself and succeed on its own merits, and (2) they underestimate the potential hostility of the environment.

- Revolutionary products can easily be derailed by unforeseen or unforeseeable external forces. But what often comes as the biggest shock for innovators is *internal* opposition from the very organization that stands to benefit most from the breakthrough.

- Adopting the ALIEN perspective helps you to think about managing risk *and* gaining traction among key stakeholders.

- ALIEN navigation strategies are built around the need to *survive* and the need to *thrive*.

- To navigate safely, ALIEN thinkers map the sources of friction and then work to neutralize them in two ways: (1) by using camouflage to evade detection and the destructive judgment of others, especially during the project's early stages; and (2) by strengthening their ability to absorb hits from critics and opponents.

- One way to minimize attacks on your project is by stressing its links with the past. Another way is to emphasize how the idea aligns with the organization's core values and goals.

- Because you can't anticipate and circumvent every threat, you need to adapt—as Owlet's founders did after they encountered opposition from hospital administrators and the FDA.

- To thrive, you need to identify positive forces to help your idea gain traction. You must also find original ways of harnessing those forces to ensure that everyone benefits—as TerraCycle did when it shifted its focus to all the untapped waste materials.

- You can also thrive by seizing unexpected and overlooked opportunities. KidZania cofounder López did this by deciding to expand into Asia *before* tackling the US market.

- When the obvious partners turn you down, ALIEN thinking helps you identify other candidates. After Bertrand Piccard found that the aviation industry had zero interest in building a solar plane, he turned to yacht builders.

QUESTIONS TO ASK YOURSELF

1. What are the forces that can make or break your solution?
2. Your solution may be extremely innovative and promising, but are you aware of how it might encroach on other people's turf internally and externally? Do you understand all of the frictions it could create?
3. To make your disruptive idea less threatening, are there ways you could link it to the organization's past heritage? Can you show how it supports the achievement of a mission that is highly relevant to your key stakeholders?
4. Recall moments when you failed to rally key decision-makers behind your cause. Are there lessons you can draw that can inform how you'll secure buy-in on your next innovative project?
5. Do you have a plan B if your idea meets with internal resistance? Can you foresee multiple paths to take until you can finally declare victory?

On Digital

1. Can you use online services (like Ideafoster) to test responses and get feedback on new ideas?
2. Can you use data to help form arguments that will convince stakeholders of the power of your ideas?
3. Have you sought advice about your ideas on websites like Quora or Reddit?
4. Have you used collaboration networks—for example, corporate collaboration networks or social media—to promote and socialize your ideas?
5. Have you tried to build viral support for your ideas?
6. Have you used psychographics to influence key stakeholders about the power of your ideas?
7. Have you tried implementing reverse mentoring so that senior leaders become more digitally savvy?

SEVEN

The ALIEN Model in Action

ALIEN THINKING IS NOT SOMETHING YOU DO AT A PARTICULAR time, in a dedicated space, with clear rules and staple props like whiteboards and sticky notes. It's not a set of gimmicks but a mindset change to achieve creativity and to turn ideas into solutions. ALIEN thinking is something you can call on at any time, whenever you hit a roadblock. It's something you have on tap when needed.

Of course, it's difficult to use an innovation model you can't keep in your head, so the ALIEN thinking model was conceived for ease of recall in two ways: as a goal and as a methodology.

First, the metaphor serves as a useful reminder of the mindset needed to generate and develop original ideas. It captures the mix of curiosity, alertness, questioning, probing, and learning associated with a state of creativity and with finding yourself in novel surroundings. The word *alien* symbolizes the quest to approach problems or opportunities from a fresh perspective, not as an expert but rather as an explorer—showing respect for facts but irreverence for "knowledge."

Second, as an acronym, "ALIEN" serves to help you memorize the five core challenges of the innovation process. The previous chapters focused on each challenge in turn, showing how they help you fight orthodoxy and elevate your innovation capability. In the next two chapters, we will pull them together and illustrate how they complement each other.

Shipping is an archetypal example of a traditional activity revolutionized by ALIEN thinking. It serves as a useful illustration of all five keys to breakthrough solutions.

THINKING IN AND OUTSIDE THE BOX

Back in the early 1950s, Malcom McLean, a onetime truck driver, was running a successful trucking company. But he could see the rising cost of highway congestion and road transportation regulations on his business—as well as the risk that domestic shipping firms, with cheap access to war-surplus cargo ships, might undercut his trucking business. His is one of the most comprehensive and deliberate examples we've found of ALIEN thinking at work.

A. Turning his attention to the shipping lines that ran along the US coastline, he noticed that ships spent more time loading and unloading a myriad of crates, barrels, and bags than they did sailing.

L. Lifting his thinking—levitating—he took time to make sense of this observation and to integrate it with his previous experience as a frustrated truck driver, waiting in line all day for his bales of cotton to be unloaded at the dock. He concluded that cheaper shipping did not require faster ships so much as the smarter loading of ships. Although he had no shipping experience, he suspected there were ways to improve the inefficient transfer process between trucks and ships. In effect, he reframed a competitive threat as a potential opportunity.

I. McLean imagined building terminals that would allow his trucks to drive up ramps and deposit the trailer section on specially designed ships.[1] Ships could do the bulk of the long-distance hauling, with trucks at either end to pick up and deliver to individual clients in one integrated transportation network.

E. After experimenting with this approach, McLean realized that each trailer would waste a lot of valuable shipboard space. So he pivoted. He proposed loading just the trailer box, not the chassis, onto the ships. The big advantage was that containers, unlike trailers, could theoretically be stacked, allowing each ship to carry far more cargo. Further experimentation went into developing containers that were both strong and light, that could be clipped and unclipped easily, and that were stackable and lockable.

Of course, the containers were just one part of the puzzle. Suitable cranes and ships did not exist, so McLean hired engineers and naval architects to design equipment capable of handling the heavy containers. He bought an aging oil tanker, the *Ideal* X, and converted it to carry fifty-eight of his newly designed containers. The test journey, in 1956, lasted five days. The cargo took less than eight hours to load rather than several days, and the cost per ton was less than one-thirtieth of the cost of hand-loading a ship.

N. McLean then had to navigate a maze of social obstacles that made inventing the equipment look like the easy part. To make this a viable solution, he needed access to ports. In spite of lobbying from the railway and trucking industries, McLean managed to acquire the Alabama-based Pan-Atlantic Steamship Corporation, not for its vessels but for its shipping and docking rights in prime eastern port cities.

Next, he needed to convince the port authorities in those strategic hubs to develop terminals with container cranes. His big break was persuading New York's Port Authority to dedicate the loss-generating New Jersey side of the harbor to container shipping. Its directors were so taken with his vision that they became early and vocal supporters of containerization.

McLean also had to contend with resistance from the powerful dockworkers' unions, which foresaw that the shift to container freight would eliminate thousands of jobs. But ultimately, the huge financial savings generated by the changes facilitated severance agreements, as well as reviving the economic fortunes of the port cities that adopted containerization.

McLean's efforts fundamentally transformed the centuries-old shipping industry. By developing the first safe, reliable, and cost-effective approach to transporting containerized cargo, he ushered in the era of mass global trade.

The ALIEN thinking model helps demonstrate how a single person can conceive and drive an idea through the full innovation cycle. McLean's example also echoes our message in Chapter 4—that creative breakthroughs often come from what researchers call *adjacent outsiders* who have enough grasp of a problem to transfer their novel perspective to solving it.[2]

SHIPPING'S NEXT BIG IDEA

Since McLean pioneered the first container vessels, shipping companies have vastly improved the speed, size, and efficiency of vessels. However, they have not altered the way cargo is carried. The sector has not had a real shake-up in six decades. Copenhagen-based Maersk, now the dominant player in the field, is trying to change that. Three years ago, its executive team realized that it needed to promote radical innovation. As Paolo Tonon, its head of maritime technology, noted, "Unconventional thinking is required if we are to make a difference in the commoditized market we are in."[3]

The top team at Maersk realized that the existing approach was producing plenty of incremental innovations *at the expense* of early-stage ideas with potentially greater impact. Maersk desperately needed to self-disrupt. Although the emphasis of the challenge differs for incumbents and external entrepreneurs, the ALIEN thinking model holds for both.

Established players may have the clout and resources to scan widely, develop or acquire ideas, experiment thoroughly, and navigate obstacles. However, they often have trouble seeing the world as it *could* be—or imagining radical departures from their business model.

Constrained by their existing worldview, they struggle to think in unorthodox ways and to recognize the ideas in their midst, especially when these ideas come from the fringes or require collaboration across silos. Good ideas get killed by internal politics when the corporate immune system kicks in. They also get killed by processes that make good ideas look like bad ideas by measuring them against proposals with clearer business cases and shorter payback horizons.

This is not just a problem for Maersk. As engineering professor Frido Smulders, who worked with Unilever to improve its research and development, noted, "Unilever is actually really good at innovation. The problem is that, over the years, the innovation process has become predictable. Innovation is associated with a certain degree of unpredictability—adventurous, disruptive innovation. But that simply doesn't make it through the pipeline with all the criteria—i.e., the stage gate or funnel—as they are currently organized at Unilever. Unilever would like to change this in order to create more room for disruptive innovation."[4]

It's a common problem for many dominant players. In some ways, the bigger your organization, the smaller your world.

Maersk's innovation output was stifled by its own system. To combat these dysfunctions and augment breakthrough thinking, Maersk essentially applied ALIEN thinking to fix the innovation process.

A. To sharpen attention, Maersk revised the mandate of the Group Innovation Board (GIP), shifting its focus to early-stage ideas and stressing the importance of looking for ideas outside as well as inside. In close partnership with the executive board, the GIP met every two months to prioritize initiatives and allocate spending to the most strategically relevant projects.

L. To support levitation, Maersk organized an annual innovation symposium to inspire meta-thinking about innovation. It also created workshops, run by the twenty-member Ideation

Community, to share best practices for generating, refining, and nurturing new ideas.

I. To spur imagination, Maersk created an innovation services team to coordinate the innovation communities across business units and to provide them with expertise from within or outside the group, as needed. It also set up a site for employees to share ideas within different innovation themes. And it started running forty-eight-hour hackathons that are open to outsiders and offer prize money, to generate game-changing ideas.[5]

E. To facilitate experimentation, Maersk gave the GIP a budget of $10 million to test and validate ideas that are difficult to manage alongside daily operations and require a longer payback time. Maersk also teamed up with accelerators in Silicon Valley (Plug and Play Tech Center) and Copenhagen (Accelerace) to help it connect with start-ups that might want to run a pilot project with Maersk.

N. To improve navigation, Maersk invested in the creation of a toolbox to improve how it works across business lines and with start-ups. It appointed an innovation portfolio chief, Anneli Bartholdy, to facilitate collaboration. She explained, "We have a very different working culture, working speed, bureaucratic process, to the [partner] start-ups. . . . So we are figuring out what structure you need to have in place, how you enable and support these sorts of activities, and work in a different way, whether it's with a start-up or just with different sorts of projects than we're used to."[6]

An early result of the revised innovation system made headlines when Maersk used a drone to deliver a bucket of cookies to a tanker at sea. This was the first test of a much bigger idea.

A diverse group of ship builders, naval architects, and innovators from outside the marine industry—including a rocket builder—were challenged to imagine the future of container shipping. They came up with the concept of a ship that never docks.[7] It's an unmanned, radio-controlled ship that stays at sea while a flock of drones pick up the containers and deposit them safely onshore.

This radical idea questions the very need for ships to come into port. (By the way, this is where the term *opportunity* actually originated. The Latin phrase *ob portu* describes the right wind and tide conditions needed to carry sailing vessels safely into port.) Fittingly, the Maersk team found opportunity by ignoring a basic assumption.

Like McLean, the Maersk team was thinking in terms of moving goods from A to B rather than fixating on the mode of transportation. This line of thinking may dramatically change the entire industry, but for now, it remains an intriguing idea. As with McLean's idea for containerized shipping, it will take much more ALIEN thinking, particularly experimentation and navigation, before the notion becomes a solution.

THE MISSING INGREDIENTS

By digging into the back stories of multiple cases and deconstructing the process used by change makers to develop breakthrough solutions, we noted two key inconsistencies in how innovation is described by established innovation models. These models systematically underplay two aspects of the innovation process, the ones we call levitation and navigation. The first is conspicuous by its absence, while the second is misrepresented as a mere formality.

We can illustrate this by returning to the containerization example.

Levitation

Existing innovation models neglect the importance of creating the space (or taking a time-out) to process and integrate disparate information in unexpected ways.

In a fast-paced world, a willingness to step back and pause, to rethink and synthesize new inputs, becomes all the more important.

It provides a competitive edge over rivals who rush headlong toward incremental solutions.

During the creative process, you periodically need downtime from the activity to revitalize your understanding: to make sense of the cues you have noticed and, sometimes, to let your subconscious mind carry on working. What sleep is to the mind and body, pause is to innovation.[8] Individually or collectively, you have to disengage to reflect on what you are innovating and *how* you are innovating. Again, levitation is the engine that guides the other four strategies and prompts switches between them.

Although the practice of reflection is implicit in most models, we believe it needs to be explicit. It was included in Graham Wallas's pioneering 1926 model of creativity as "incubation," but recent models prefer to emphasize speed and action, ignoring the role of deliberate procrastination and unconscious processes. The problem with frenetic cycles of build-measure-learn is that they leave little room for reflection to consolidate your learning or take in the big picture.

The cases of Ferran Adrià, Narayana Peesapaty, Marcus Raichle, and others show that gaining perspective is key to the innovation process. At critical junctures, each of these innovators stepped back to allow emotions, thoughts, or information to sink in, and to work out what these things meant or *could* mean.

Take the extreme case of Malcom McLean. He first noticed the inefficiency of shipping companies back in 1937, when he was just a truck driver queuing up to unload his goods. His breakthrough idea gestated for eighteen years before he finally acted on it. This is another example of the slow hunch mentioned in Chapter 4. The information took time to percolate—to coalesce with other data points—before it eventually crystallized into an idea.

In the intervening years, he absorbed a lot. He built up the means to experiment and the credibility to secure bank loans, as well as the confidence, skills, and experience that would enable him to look at the shipping industry as a whole and recognize the causes of its inefficiencies. During this period, he also came to understand the fundamental economics of any transport business, including the principle that vehicles only earn money when they are on the move. By the time he looked again at shipping, his experience had transformed his outlook and his capacity to make sense of the whole system, not just the inefficient flows between modes of transport.

Without advocating for endless reflection, we feel it's necessary to highlight this overlooked aspect of innovation. Levitation allows you to reshape the opportunity that precedes the idea—to identify the root causes, not just the symptoms.

To have real impact, you need to step back to see if you are identifying the big strategic opportunities or just strands of an opportunity. Opportunities look different to different entities or people—or even to the same person over time.

If you don't take the time to collect your thoughts, you can easily overreact to user information or feedback without working out what users are *really* telling you. You can get caught in an acceleration trap, where momentum is privileged over deliberation and where you converge prematurely on a narrow opportunity, missing out on more ambitious innovation.

Levitation is a key mechanism through which the creative process progresses. It forces you to think twice—to question assumptions and reconsider options, discrepant data, or cognitive frames. Without levitation, there is no "big I" innovation of the sort that looks beyond current horizons and takes time to gestate.

Discussing his creative inspiration, François Englert, a Nobel laureate in physics, observed, "When I was a professor at the Université Libre de Bruxelles, I bought a second-hand mattress at the flea market and put it in my office. . . . Most of my original research did not result from deductive thinking. It arose when I let my mind wander freely while lying on this mattress. If you think deductively, you only discover what you suspected at the beginning of the process. With the other method, you unleash your unconscious brain and in the end, it comes up with totally new ideas."[9]

Of course, from a digital age perspective, the opportunity for levitation is shrinking. McLean had eighteen years. Today, you're lucky to find eighteen minutes! Carving out the necessary time becomes critical.

Navigation

Existing models of innovation minimize the complexity and conflicting interests of the ecosystem that stands between you and those you hope to serve. They underplay the ingenuity needed to gain traction, within the organization and beyond, and the level of organizational reinvention sometimes needed in the final push to market.

A common pattern among these innovation models is the use of labels like *implementation, execution, scaling up,* or *launch* to designate the final stage. The problem with these terms is that they make the challenge sound straightforward—more grind than spark. They imply that the time for thinking and creativity is over and now is the time for doing.

This false separation between conceiving the solution and executing it disregards the creativity and cleverness needed to achieve internal buy-in and engage intermediaries, partners, and other stakeholders in unexpected ways. Misrepresenting the challenge in this way encourages naive engagement with the ecosystem, suggesting that once you have a superior offering, the rest is smooth sailing.

For this reason, we prefer the term *navigation*, which helps convey the planning, resourcefulness, and improvisation needed to negotiate uncharted waters and steer a solution to a successful conclusion.

Originality is required throughout the innovation process. The last stretch offers as much scope for inventiveness as any of the other four pursuits. Again, this is well illustrated by the case of Malcom McLean. He excelled at navigation.

His breakthrough thinking was not so much the reinvention of the container, the ship, or dockside crane. Similar technologies already existed for rail and sea links. The *real* leap lay in the profound system redesign that he instigated.

Two turning points are especially worth highlighting. First, the trucking companies, shipping companies, and ports, with their divergent needs, demanded containers of different sizes. A giant stride was getting the various players to agree on standard dimensions so that any container could fit on any ship and be handled by a crane in every port. Second, McLean decided to allow the industry to use his company's patents royalty-free. That way, every container in every country could use the same corner fittings.

These breakthroughs cleared the way for container shipping to become a global business. And they relied on McLean's outsider status, his unorthodox approach, and his skill in playing the system. To prevail, he had to show the different stakeholders that his ALIEN proposal was not contrary to their vested interests.

The ALIEN thinking model specifically acknowledges the critical role of levitation in shaping the journey and navigation in negotiating the many barriers to implementation.

KEY TAKEAWAYS

- ALIEN thinking is not something you do at a particular time, in a dedicated space, with clear rules and staple props like whiteboards and sticky notes. It's something you can call on at any time, whenever you hit a roadblock, to turn ideas into solutions.
- Malcom McLean's transformation of the shipping industry via containers demonstrates how a single person can conceive and drive an idea through the full innovation cycle.
- The ALIEN thinking challenge differs for established players and entrepreneurs. Incumbents have the resources to scan widely, develop or acquire ideas, experiment thoroughly, and navigate obstacles, but they often struggle to imagine radical departures from their current activities.
- For established companies, ALIEN thinking can help to deliver a particular breakthrough or it can be applied to the innovation process itself—as Maersk did in order to promote more radical innovation.

- Existing frameworks underplay two key aspects of innovation: the critical role of reflection (levitation) and the need for creative steering (navigation). The first is conspicuous by its absence, while the second is misrepresented as a formality—mere implementation.

TABLE 7.1 OVERVIEW: ALIEN STRATEGIES

STRATEGIES	PURPOSE ("WHY")	OBJECTIVES ("WHAT")	TACTICS ("HOW")
Attention			

See the world with fresh eyes to perceive reality as it is	To avoid a clever solution to a problem not worth solving	• Appreciate emerging trends
		• Spot anomalies and weak signals
		• Identify problems worth solving
			Improve and challenge your vision
			
			1. To see better
			  • Zoom in
			  • Zoom out
			2. To see differently
			  • Switch focus

Levitation			

Step back to gain perspective	To avoid a narrow solution to a larger problem	• Make sense of what you've learned
		• Reflect on what matters most
		• Reframe the problems you want to solve
			Redirect your efforts
			
			3. Time-out (brief interlude)
			  • Figure out *what* you think
			  • Review *how* you think
			4. Time off (change of activity)
			  • Oxygenate your mind
			  • Take a mental break

Imagination			

Defy convention to produce out-of-this-world ideas	To avoid a conventional solution to a complex problem	• Question the status quo
		• Look beyond obvious solutions
		• Steal with pride from other domains
			Envision that which is not
			
			5. Release your imagination
			  • Adopt a playful state of mind
			  • Brainstorm questions
			6. Stimulate your imagination
			  • Use analogies
			  • Combine concepts

Experimentation			

Test smarter to learn faster and cheaper	To avoid an undercooked solution to a real problem	• Test without building
		• Discover what works, but stay flexible
		• Make every failure count
			Explore options and test assumptions
			
			7. Welcome surprises
			  • Propose multiple models
			  • Provoke extreme reactions
			8. Accept surprises
			  • Let the data speak
			  • Seek people who think differently

Navigation			
Maneuver to soar and avoid being shot down	To avoid a great solution that fails to make an impact	· Identify the value-critical stakeholders · Look for unconventional partners · Bridge conflicting interests	**Anticipate and adapt to the forces that can make or break your solution** 9. Survive · Map frictions · Neutralize threats · Resist attacks 10. Thrive · Map opportunities · Leverage opportunities · Seize unexpected opportunities

TABLE 7.2 ALIEN MOVES: WHAT YOU CAN DO DIFFERENTLY

From our research and discussions, here are some simple pointers (by no means exhaustive) to get you started.

STRATEGIES & TACTICS	TIPS & TRICKS
Attention *See the world with fresh eyes* *To see better* · Zoom in · Zoom out *To see differently* · Switch focus	· Acknowledge your professional bias · Look from multiple angles · Look from the inside, as a participant, not just an observer · Watch for what people *don't* do or say · Survey from a higher vantage point—i.e., take a bird's-eye view · Search beyond the usual suspects · Look at peripheral groups, such as extreme users (*ultras*) · Spend time with people on the fringes · Look from the opposite angle—e.g., switch from extreme users to non-users · Take on an alternative persona—e.g., consider the world from the perspective of a child or a blind person
Levitation *Step back to gain perspective* *Time-out (brief interlude)* · Figure out *what* you think · Review *how* you think *Time off (change of activity)* · Oxygenate your mind · Take a mental break	· Pause to make sense of what you've observed · Ask yourself what's missing · Observe yourself as another would · Question the objective and the approach—should you redirect your efforts? · Change your surroundings—e.g., leave the house or office · Immerse yourself in a different body of knowledge, discipline, or profession · Read a book in an adjacent field—e.g., science fiction for a scientist · Attend a conference in a related sector · Visit a museum · Do something mindless—e.g., doodle, stare out of the window · Take a tech break—e.g., at lunch, while driving or commuting to work · Take time to think—e.g., sleep on it, take a walk without your smartphone

continues

continued

Imagination *Produce out-of-this-world ideas* *Release your imagination* • Adopt a playful state of mind • Brainstorm questions *Stimulate your imagination* • Use analogies • Combine concepts	• Play with constraints or presumed requirements—e.g., discard an imperative or add a constraint • Consider how *not* to do something—e.g., think about how Amazon/Google would not do it • Change how you describe the problem or rephrase the goal • Generate as many questions about the challenge as you can • Ask yourself why, why not, what if, and how • Ask yourself, "What if we no longer did what we do now?" • Produce as many original ideas as possible, without judgment or self-editing • Compare your problem to a parallel challenge somewhere in nature • Explain to someone unfamiliar with the problem why you are stuck • Tap into a Reddit community and follow the discussion thread in unlikely directions • Talk to outsiders with a different bag of experiences • Steal with pride from other sectors
Experimentation *Test smarter to learn faster and cheaper* *Welcome surprises* • Propose multiple models • Provoke extreme reactions *Accept surprises* • Let the data speak • Seek people who think differently	• Test ways to acquire faster, richer, and more unexpected data from the outside world • Look for natural experiments • Use technology to test multiple ideas in a frugal way • Share your "shitty first draft," ugly mock-up, or Frankenstein prototype to start a dialogue • Try out your offering on early adopters—those willing to make sacrifices to get their hands on it—to create an iterative feedback loop • Stress-test in harsh environments • Seek perspectives that really stretch your thinking • Collect feedback from "tough crowds" and take it seriously • Surround yourself with people willing to level with you and capable of challenging your interpretations of the evidence
Navigation *Maneuver to soar and avoid being shot down* *Survive* • Map frictions • Neutralize threats • Resist attacks *Thrive* • Map opportunities • Leverage opportunities • Seize unexpected opportunities	• Size up the key forces liable to make or break your solution • Identify and prioritize the most important ecosystem players • Get inside help to steer through the corporate immune system • Assess who has the most to gain or lose • Picture yourself as the enemy • Role-play the reactions of critical stakeholders • Stay low until you've got something to show • Buy yourself time and space, scrounge resources, don't make yourself a target • Win the support of champions who can provide you with cover • Craft different messages that resonate with backers and blockers • Build bridges by talking their language • Stress the continuity of the solution more than its disruption • Link your radical project with familiar and accepted concepts • Devote as much energy and creativity to developing the business model, structure, and processes as you do to the product • Consider unconventional partners to outmaneuver blockers • Be ready to swerve or shift into reverse

EIGHT

A Flexible Sequence

FOR THE SAKE OF CONVENIENCE, WE HAVE PRESENTED THE FIVE elements of ALIEN thinking chronologically. In practice, though, they constitute not an orderly sequence or even a cycle but a mix that requires you to crisscross between activities.

Conventional creativity and innovation models—such as the stage-gate model or the waterfall model—don't stress this enough. Instead, they portray the pursuit of novel solutions as a tidy progression through successive steps, with review points at the end of each phase.[1] Even design thinking can be too linear and rigid in its insistence on starting with users.[2]

In practice, the process of discovering breakthrough solutions is more fluid. Initiatives rarely proceed as expected. Innovation is inherently unpredictable. A comprehensive innovation model must reflect these realities. Otherwise, it will constrain your creativity or lead you astray. Our ALIEN thinking model accounts for two fundamental realities often overlooked by established innovation methodologies. We are agnostic about where to start and which path to follow (see Figure 8.1).

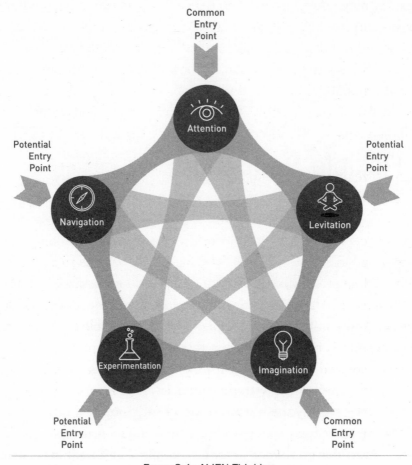

FIGURE 8.1: ALIEN Thinking

MULTIPLE ENTRY POINTS

Although attention may be a logical starting point for innovation, beginning with the others is perfectly valid, too.

Levitation sometimes comes first. Your attention may be captured by something precisely *because* you have stepped away. Taking time off can make you more attentive to idiosyncrasies you might not otherwise notice. For example, Clarence Birdseye, the frozen food pioneer, never set out to revolutionize food preservation. In 1916, while on assignment in a sub-Arctic part of Canada, the assistant naturalist would go ice fishing with the Indigenous Inuit people. With time on his hands in this alien environment, he became receptive to their practices, learning how to ice fish and freeze the day's catch like the Inuit. He was surprised by how incredibly fresh the frozen fish tasted once thawed and wondered how this could be. Five years later, now working for the US Fisheries Association, he observed the problem of getting fresh fish to market in good condition and started experimenting with freezing fish so that it retained its appearance, taste, and texture. The trick, he realized, was speedy "flash-freezing." But it took three more years to invent the machinery, facilities, and distribution infrastructure to capitalize on that idea and create the frozen food industry.[3]

Imagination is another possible entry point. When Argentine car mechanic Jorge Odón dreamed up his medical device as an alternative to forceps deliveries, he was not consciously looking to improve on existing birthing equipment. The idea literally came to him in his sleep. At 4 a.m., he woke his wife to share his epiphany. His wife, he recalled, "said I was crazy and went back to sleep."[4] The next day, he asked a skeptical friend for an introduction to an obstetrician, and the long journey began. Eureka moments are overhyped, but they *do* exist. You sometimes come up with an idea you weren't looking for—perhaps to a problem you don't even have.

Arguably, imagination was also the entry point for Bertrand Piccard's solar plane. This example shows that to achieve a breakthrough solution, you don't necessarily have to know something no one else does. You can get there by *believing* in something no one else does. As Piccard recently observed, "To start something like that you have to be a little bit naive at the beginning. You have to ignore all the problems you'll have later."[5] Existing models often have difficulty accommodating this kind of big-leap innovation, which is based largely on a top-down belief in possibility rather than on current needs or technologies. Part of our genius as humans is imagining that which is currently out of reach. "This is the fate of pioneers," said Piccard. "In the beginning, people think it's impossible, it's stupid, it's ridiculous. Then it happens and it becomes obvious."[6]

Experimentation can be the gateway to breakthroughs when you stumble upon a finding that is unrelated to your focus of inquiry—leading you not merely to pivot but to reboot and set off in another direction altogether. This is not uncommon with scientific discoveries that originate as the unintended by-products of other research—as in the classic case of the blockbuster drug Viagra, which was the accidental result of Pfizer's clinical trials for new heart medication.

More recently, Jeannette Garcia, a chemist at IBM Research who was seeking to synthesize a polymer, set up a chemical reaction and then stepped away to fetch an ingredient. Returning to the heated flask, she found a bone-hard substance. It turned out she had discovered a new family of recyclable thermoset polymers—the first new class of polymers in decades.[7] The new substance was superstrong, lightweight, and, unlike comparable materials, easily and infinitely recyclable. This unique combination of properties made it a breakthrough discovery with a wide range of potential applications in aerospace, autos, electronics, and 3D printing. But for now, it remains a solution in search of meaningful problems.

Navigation can also be the first step in the process, as when two or more parties decide to create something together. This was famously the case for design students Brian Chesky and Joe Gebbia. Chesky gave up a secure job to move to San Francisco with the aim of building a business with his friend, which he assumed would be a design company. Sharing a three-bedroom apartment, they struggled to cover the rent and decided to make up the shortfall by renting out three air mattresses to people attending an upcoming international design conference. They cheekily advertised it as "Airbed and Breakfast." This is how Airbnb, a business now worth over $38 billion, was born.[8] Sometimes the innovation process starts because you set up favorable creative conditions, without necessarily responding to a perceived need.

MULTIPLE PATHWAYS

Just as you can start anywhere in the creative process, you can also proceed in any direction and then switch focus as required. Hardly any innovation models explicitly acknowledge such freedom. For this reason, they are often taken too literally and reduced to rigid recipes that produce disappointing outcomes.

In practice, innovation proceeds in a more flexible, nonlinear fashion, and the ALIEN thinking model is more accepting of this essential messiness. We recognize that the five activities can affect each other in unpredictable ways—or even occur concurrently. It's more like switching between lateral thought processes than following a cycle or a sequence.

Consider again Jeannette Garcia's polymer discovery. Having created something unexpected, she then had to pay attention. Had she not been receptive to new outcomes and curious about her mistake, she could easily have dismissed it as a failed experiment. Instead she returned to experimentation, joining forces with IBM's computational chemistry team to work back from the final polymer and discover the causes of the surprise reaction. Looking ahead, IBM has twin priorities to juggle: It must define relevant customer needs (attention) while at the same time deciding with whom it must partner (navigation) to bring a solution to fruition. This may require a great deal of back-and-forth, along with further experimentation (to produce prototypes) and occasional levitation (to reevaluate and avoid overstretching).

The creative process is full of twists and loops that will require you to revisit your original question, the options for responding, or the choice of partners.

Knowledge acquired from experimentation may lead you back to attention to reconsider the needs of users or back to imagination to generate alternative ideas. Sometimes you will address the challenge of "asking the right people the right questions" late in the process—as when Owlet realized that wireless health monitors were a pressing need, not for hospital decision-makers, but for anxious new parents. You must follow whatever sequence works for you. Process fixedness is a barrier to breakthrough solutions.

According to Swatch creator and serial entrepreneur Elmar Mock, "Innovation is like a labyrinth with a lot of dead ends and a lot of possibilities. You know what you want to start to do but you don't know where you'll end it. To reach the target is not to reach the point you defined at the beginning. It is to reach the right exit that can bring you to success."[9]

An approach designed to stimulate radical thinking cannot be tightly regimented. To illustrate how the innovation process can diverge from the expected trajectory, consider the problem of land mine removal.

ZIGZAGGING ALL THE WAY

Abandoned land mines kill and maim thousands of civilians each year. They also deprive communities of access routes to villages, water, and roads. Removing land mines is a dangerous, tedious, and costly process, requiring specialized demining vehicles, operators with metal detectors, or highly trained dogs and handlers. But Bart Weetjens, a former product engineer from Belgium, came up with a program that is better, faster, and cheaper than existing solutions.

It all started with **levitation**. Frustrated with designing buses for a living, Weetjens decided to quit his job in 1995 to find a more satisfying way of contributing to the world.

Jobless for several months, he dabbled as an artist, pondering which path to follow next, until his **attention** was drawn to the issue of land mines through the campaigning work of Diana, Princess of Wales.[10] So he traveled to Mozambique and Angola to look at the problem from the point of view of a subsistence farmer in an African village who was unable to access his farmland because it was filled with land mines.

Pulling back to contemplate the problem from a different vantage point, he chose to attend a seminar on land mine detection systems (**levitation** again). There he learned of a stalled 1970s study in which gerbils had been taught to recognize the scent of explosives through electrical brain stimulation.

This finding set his **imagination** racing. Having bred rodents as a teenager, Weetjens wondered whether rats, a less nervous animal with an equally keen sense of smell, could be trained to detect land mines. But when he approached the Belgian military and other organizations with the idea of using rats in demining operations, they scoffed at his plan.[11]

It took him two and half years to finally convince professors at the University of Antwerp to take him seriously and help him secure a one-year grant—enough to start a rat-breeding program in Belgium. Note that in this case, a lengthy spell of **navigation**—convincing key supporters and funders—preceded any form of **experimentation**.

He then conducted several rounds of **experimentation** to develop a robust training program and identify the best type of rat for the task. The first batch of rats imported from Africa proved very territorial and bit each other to death. Even as they experimented in the lab, Weetjens continued to pay **attention** to the conditions on the ground and to reframe what the program needed (**levitation**). For example, the rats would have to be big enough for people to see them in the vegetation, but not so big that they set off land mines. Also, the rats needed to have a life span that was long enough to provide a good return on the training investment. After consulting rodent scholars, Weetjens eventually settled on the African giant pouched rat, which could live in captivity for up to eight years.

Conditioning the rats required **experimentation** with various training protocols to identify the best behavioral reinforcement. Weetjens and his team finally settled on clicker training, in which the animal learns to associate smell recognition with a clicker sound and a food treat. After two years of preliminary studies, there was sufficient evidence that the training worked to transfer the program and the research team to Africa to pursue training with local handlers.

Of course, this raised new **navigation** challenges. First, there was the choice of country. Among multiple candidates, Weetjens settled on Tanzania, which had the right combination of commitment to the project and the necessary political stability. Weetjens set up a broad collaboration agreement with the country's defense forces for logistical support and partnered with the Sokoine University of Agriculture in the eastern part of the country. The latter donated a plot of land on which Weetjens's NGO, now called Apopo, established its central research and training facility.

Second, in terms of funding, Weetjens continued to look for donors in the international community. A list of partners that originally included only the University of Antwerp and the Belgian government grew to include thirty groups, some of which had originally dismissed his idea.

Third, Weetjens had to overcome resistance within the local communities. Rats are widely seen as pests in Africa, resented for the damage they do to crops and food stocks. To help change those preconceptions, Weetjens dubbed his animals "HeroRATS"—and once local populations saw what the rats and their handlers could do, their repulsion turned to curiosity and, ultimately, affection.[12]

Reflecting on the reversal in attitudes (**levitation**), Weetjens thought it might be possible to leverage the paradoxical image of the rat as savior for fundraising purposes. He launched a virtual adopt-a-rat program to encourage private individuals to sponsor training and care (**experimentation**).

The reality of breakthrough solutions is that the five ALIEN thinker strategies all bleed into one another. Breakthrough solutions do not follow a standard recipe: they are complex and happen in a different way each time.

For Apopo, the results have been spectacular. Within a year of starting, an external evaluation by the Geneva International Centre for Humanitarian Demining (GICHD) confirmed the efficacy of the program on multiple fronts.

A specialist with a mine detector can sweep an area the size of a tennis court in about two days. A trained rat can do it in thirty minutes.[13] Unlike dogs, rats are not troubled by the withering heat, and they are less prone to tropical diseases. Their training costs roughly a third of what it costs to train a dog, and they are easier to transport and look after. And unlike dogs, they do not bond to individual trainers and are not prone to ennui.

Compared to demining vehicles, rats are cheap and widely available. They offer a local, sustainable, and scalable solution to help disadvantaged communities tackle the land mine problem more independently, rather than perpetually relying on imported know-how or technology. They are now deployed in other African countries, including Mozambique, Angola, and Ethiopia, as well as in South America (Colombia) and Asia (Cambodia, Laos, and Vietnam).

But the story doesn't end there. The opportunity afforded by this low-tech solution turned out to be much broader than Weetjens originally envisaged.

SMELLS LIKE INNOVATION

On a long-distance flight (literal **levitation**) to the operational headquarters in Tanzania, Weetjens found fresh inspiration. He chanced upon a BBC news report predicting a 400 percent increase in tuberculosis deaths worldwide. It was not a problem to which he had previously paid much **attention**. But some of Apopo's staff were infected with HIV, which weakened their immune systems and made them more vulnerable to TB. As Weetjens reflected on the disease, it struck him that in his native language, Dutch, TB was called *tering*, which refers to "the smell of tar." (TB sufferers exude a tar-like smell in the very late stages of the disease.) At this point, Weetjens's **imagination** connected the dots: if TB had a smell, then rats could surely be trained to detect it at an early stage of infection. (This is also an interesting example of overcoming functional fixedness, as discussed in Chapter 4, where the fact that he had developed a "technology" for detecting land mines did not prevent him from seeing ways of extending that solution to other domains.)

So Weetjens began **experimentation** to adapt his ingenious solution to fight another deadly killer—this one counting its victims in the millions, not the thousands. According to WHO figures, nine million people fall sick with TB every year, and lab tests misdiagnose one-third of cases. In the process of experimentation, Apopo developed an automated training cage to speed up the training of rats and remove any human bias from the testing.

As it turned out, the rodents proved more adept at detecting TB than lab technicians equipped with simple light microscopes. A lab technician would need up to a day to screen fifty samples for TB, but a trained rat can do the same work in less than ten minutes.[14] As of 2019, rats had identified over eleven thousand TB cases initially missed by lab tests, increasing detection rates by 40 percent.[15]

The key challenge has been **navigation** related. Some of the people living in urban slums or without a fixed address were difficult to contact to inform them of the results. But most of them did have a mobile phone, and that was how they were tracked down and enrolled in treatment.[16]

Still, at the time of writing, the **experimentation** is far from over. Weetjens is currently working to repurpose his low-tech solution to address other latent needs. Rats are being trained to pinpoint gas leaks, narcotics, and tainted food. One day, rodents may even be seen on leashes sniffing at luggage or working in hospitals. According to Tim Edwards, Apopo's head of training and behavioral research, "Currently, there's a lot of interest in sniffing for cancer. We've also been contacted about hypoglycemia and some other medical applications. There's so much potential; it's just a matter of finding the time and the resources to investigate it."[17]

The rats' natural burrowing talent can even be harnessed to search for survivors in the rubble of collapsed buildings—by fitting the rats with miniature wireless cameras on their backs. Again, **levitation** is needed to work out which opportunities to prioritize.

So what initially looked like a quirky alternative to clearing land mines turned into a generic solution for multiple problems with a "smellprint"—in much the same way that TerraCycle's recycling solution for waste food became a model for more complex types of waste (see Chapter 6). Besides illustrating the convoluted pathways to breakthroughs, the case of Apopo underscores the power of ALIEN thinking to challenge your perceptions of the resources at your disposal. You sometimes have more than you think.

THE RECIPE FALLACY

The Apopo case shows how the five drivers of ALIEN thinking—attention, levitation, imagination, experimentation, and navigation—

pop up repeatedly throughout the process, but not necessarily in that order. While one strategy may dominate at any given time, they actually interweave and act as relays between each other. The "stages" may be reversed, repeated, or conducted in parallel. The combinations are endless, but certain patterns are more common. For example:

Attention is needed throughout the process to decode the evolving needs, not just of end users, but of all internal and external players who may support or oppose your solution (navigation). You may also return to attention after reframing the problem through levitation or each time your experimentation produces unexpected outcomes.

Levitation is a recurrent concern for revitalizing your understanding and ensuring that you stay on course. After imagination, levitation helps you figure out the best opportunity—ideas that seem brilliant when they're hatched sometimes benefit from a cooling-off period—and after experimentation, it helps you integrate new data and assess whether or how much to pivot.

Imagination is vital not just to generate product and service ideas but also to find alternative ways of looking at, and thinking about, the problem, or ways of testing and delivering the solution. Imagination pushes you to ask why and what-if questions at all times.

Experimentation is key to sensemaking. We come to know the world through the cycle of taking action and paying attention to the outcomes. Experimentation delivers the feedback you need to reorient your attention, to change how you generate ideas, and to anticipate how others may react to your breakthrough solution.

Navigation is needed to anticipate or adapt to threats and opportunities (before and after you develop a workable solution), to secure the space and support to start the project, to connect with potential cocreators, and to rebuild your credibility after experimentation setbacks.

Having emphasized the elasticity of the ALIEN thinking sequence, we must add a caveat.

Although the order is flexible, you do need to touch each base at least once, because different elements neutralize different biases. Each element adds value. By attending to all five touchpoints, you will maximize your chances of reaching a truly game-changing solution at the journey's end. Disregarding even one can lead you to focus on the wrong problem, idea, or solution. For example, if you neglect the attention component, you may end up with a great solution to a problem not worth solving (the Segway). If you underinvest in levitation or imagination, then you are going to converge prematurely on a solution that may not be very original or appropriate (Bic for Her pens[18]). A lack of experimentation will lead to a rushed offering that does not meet the identified need (Theranos), while poor navigation of the ecosystem may result in a great solution's failing to have the impact it deserves (Steve Sasson's first digital camera).

In a fast-changing environment, you must be able to change the way you approach problems and think. Radical innovation cannot be preordained or reduced to a fixed sequence. If your progress from problem to solution is too orderly, you are probably curtailing your creativity. To produce novel results, you need to be able to switch between tactics. ALIEN thinking can be reconfigured without losing its integrity.

The five activities interlink in unpredictable ways. A better metaphor than a cycle or sequence is a lattice, where the five strategies run parallel and continually crisscross.

A REALISTIC UMBRELLA

By recognizing the importance of flexible strategies, the ALIEN thinking model serves as a more realistic guide to developing breakthrough solutions than other methodologies. It accepts that the innovation process is convoluted and that creativity is a journey of sensemaking.

The elements of the ALIEN thinking model are not unique—and we suggest further resources for each of the five strategies below—but collectively, they capture the full scope and contortions of the innovation process. The model specifically acknowledges the critical role of reflection in defining problems or opportunities, reorienting experimentation, and creatively negotiating the barriers to breakthrough solutions. It covers both digital and traditional approaches, and it provides an umbrella framework that can easily accommodate design thinking, lean start-up, business model canvas, and other innovation strategies.

There is no shortage of books offering advice on breakthrough solutions, but the advice comes from multiple streams. As a result, it can be difficult to reconcile the competing tools, methodologies, and recommendations. The ALIEN thinking model covers the entire innovation process, and it does so in a way that is easy to recall, in keeping with Albert Einstein's principle: "Everything should be made as simple as possible, but not simpler."

+>====<+

KEY TAKEAWAYS

- In practice, the process of discovering breakthrough solutions is very fluid. Although attention may be a logical starting point for innovation, beginning with the other components of ALIEN thinking is perfectly valid, too. Innovation has multiple entry points.
- Just as you can start anywhere in the creative process, you can also proceed in any direction and then switch focus as required. Jeannette Garcia's polymer discovery is one example of this.
- Bart Weetjens's development of "HeroRATS" shows how the five drivers of ALIEN thinking pop up repeatedly throughout the creative process, but not necessarily in the ALIEN order. The permutations are endless, though certain patterns are more common than others.
- Although the order is flexible, you *must* touch each base at least once, because different elements neutralize different biases. Each element adds value. Disregarding even one can lead you to focus on the wrong problem, idea, or solution.
- By recognizing the importance of flexible strategies, the ALIEN model serves as a more realistic guide to developing breakthrough

solutions than other methodologies. It accepts that the innovation process is convoluted and that creativity is a journey of sensemaking.

QUESTIONS TO ASK YOURSELF

1. Would you describe yourself as someone with a flexible personality?
2. How comfortable are you with starting anywhere in the creative process and then switching directions (zigzagging) if need be?

TABLE 8.1 FOR FURTHER READING

Attention	Podcast: Amy Webb, "How to Spot Disruption Before It Strikes," interview by Paul Michelman, *Three Big Points* (podcast), *MIT Sloan Management Review*, March 17, 2020.
	Article: Bill Taylor, "To Come Up with Better Ideas, Practice Paying Attention," *Harvard Business Review*, May 23, 2019, https://hbr.org/2019/05/to-come-up-with-better-ideas-practice-paying-attention.
	Article: Max H. Bazerman, "Becoming a First-Class Noticer," *Harvard Business Review*, July–August 2014.
	Book: James H. Gilmore, *Look: A Practical Guide for Improving Your Observational Skills* (Austin, TX: Greenleaf Book Group Press, 2016).
Levitation	Article: Emma Seppälä and Johann Berlin, "Why You Should Tell Your Team to Take a Break and Go Outside," *Harvard Business Review*, June 26, 2017, https://hbr.org/2017/06/why-you-should-tell-your-team-to-take-a-break-and-go-outside.
	Article: Thomas Wedell-Wedellsborg, "Are You Solving the Right Problems?" *Harvard Business Review*, January–February 2017.
	Book: Kevin Cashman, *The Pause Principle: Step Back to Lead Forward*, (San Francisco: Berrett-Koehler Publishers, 2012).
Imagination	Article: Martin Reeves and Jack Fuller, "We Need Imagination Now More Than Ever," *Harvard Business Review*, April 10, 2020, https://hbr.org/2020/04/we-need-imagination-now-more-than-ever.
	Article: Nathan Furr, Jeffrey H. Dyer, and Kyle Nel, "When Your Moon Shots Don't Take Off," *Harvard Business Review*, January–February 2019.
	Book: Welby Altidor, *Creative Courage: Leveraging Imagination, Collaboration, and Innovation to Create Success Beyond Your Wildest Dreams* (Hoboken, NJ: Wiley, 2017).

Experimentation	Article: Michael Luca and Max Bazerman, "Want to Make Better Decisions? Start Experimenting," *MIT Sloan Management Review*, June 4, 2020.
	Article: Julian Birkinshaw and Martine Haas, "Increase Your Return on Failure," *Harvard Business Review*, May 2016.
	Book: Stefan Thomke, *Experimentation Works: The Surprising Power of Business Experiments* (Boston: Harvard Business Review Press, 2020).
Navigation	Article: Cara Wrigley, Erez Nusem, and Karla Straker, "Implementing Design Thinking: Understanding Organizational Conditions," *California Management Review*, January 12, 2020.
	Article: Jurgen Stetter, "Four Ways to Get Your Innovation Unit to Work," *MIT Sloan Management Review*, March 29, 2019.
	Article: George Day and Gregory Shea, "Grow Faster by Changing Your Innovation Narrative," *MIT Sloan Management Review*, December 10, 2018.
	Article: Martin Ihrig and Ian MacMillan, "How to Get Ecosystem Buy-In," *Harvard Business Review*, March–April 2017.
	Book: Ryan Holiday, *The Obstacle Is the Way: The Timeless Art of Turning Trials into Triumph* (New York: Portfolio/Penguin, 2014).

NINE

Digital

The Power to Bring Breakthrough Ideas to Life

MELINDA ROLFS'S INNOVATION JOURNEY BEGAN IN 2014—THE year her attention was first drawn to charitable giving. While working at Mastercard Advisors, the company's consulting arm, she observed that "there seemed to be one major natural disaster after another. And in our offices there was one of those big screen TVs with CNN on it—and I was just constantly seeing these images of people who had so little to begin with and had lost absolutely everything."[1]

This caused her to wonder what Mastercard could do to help the organizations providing relief. After reflecting on the problem, she stepped back to get a wider view of what her company could do with the resources at its disposal.

She soon realized that the transaction data collected by Mastercard uncovered not just spending but also giving. Based on this critical insight, she narrowed her focus to better understanding the needs of charitable organizations. She saw that they lacked both access to data about donors and the ability to use it. It turned out that most charitable organizations were unsophisticated about extracting value from data and analytics. At best, they would study donation trends using large data sets (like those based on IRS information), but these data sets provided very little information that was timely or granular.

She concluded that there was a critical information gap between the data haves and have-nots. Although commercial organizations had plenty of information on consumer spending habits, hardly anyone was compiling meaningful information for the nonprofit sector.[2] Charities were not learning much about when and where to raise funds.

Rolfs decided to try to bridge the information gap. "Maybe there was a way we could leverage our data and analytics for social good," she later recalled thinking. "Perhaps we could see the trends in individual charitable giving in our data."[3]

Her breakthrough idea was that Mastercard could make the same kind of data and analytics available to nonprofits—for fundraising purposes—that it was already providing to merchants and financial institutions for commercial gain. The information would be offered free of charge, making it a form of "data philanthropy."

But to turn this concept into reality, she first needed to test it. So she entered the idea into the company's annual innovation contest. "This was a great platform for me because it forced me to think about this idea holistically."[4] Her idea won the contest, which not only helped her sell the plan to Mastercard bosses but also convinced them to let her lead the initiative.

Today, Rolfs's unit produces an annual report with a breakout for nine categories—such as education, the environment, health, and animal welfare—to help nonprofits benchmark their fundraising performance and see which causes are gaining popularity. It also generates insights on when and how giving occurs and the underlying factors that affect it. These insights serve as an anchor for donation strategies, helping nonprofits plan the timing of their campaigns and capitalize on trends in recurring donations. In addition, the program now includes external data sets on top of Mastercard's own data. (For example, political donation data has shown that charitable giving takes a significant hit in the months leading up to US elections, both presidential and midterm, because people are giving to their favorite candidates.)

Rolfs sees her ongoing challenge as reaching out to policy makers, nonprofits, academia, and the private sector to find ways of unlocking the power of data to create sustainable solutions to society's greatest problems. "The goal," said Rolfs, "is to equip organizations focused on social good with the information they need to reach more people and better serve their communities."[5]

Data philanthropy was not an idea original to Rolfs, but she effectively steered it from an attractive concept to a viable solution—one that brings value for hundreds of thousands of people. To achieve this, Rolfs adroitly applied ALIEN thinking principles. The core of her contribution was to combine the human side of philanthropy with the data and digital capabilities of Mastercard.

First, she leveraged massive quantities of information from her company's databases. Neither Rolfs nor anyone else had to physically collect the data. It was created automatically, at an immense scale, and in real time. Then she leveraged the power of analytics to generate insights without the biases that often accompany human-based analysis. Finally, she quickly built a comprehensive solution that included data from inside and outside of Mastercard. The quantities of data, strength of unbiased insights, and scale of growth could not have been achieved without a combination of human and digital capabilities.

ALIEN THINKING AMPLIFIED

As this case illustrates, combining human and digital capabilities can spur ALIEN thinking in entirely new directions and across domains. Digital technologies are helping blur the boundaries not just between commercial sectors but also between the private and public sectors, between for-profit and not-for-profit organizations, and between big businesses and entrepreneurs. The underused data that organizations routinely collect from consumers and citizens can lead to breakthrough solutions in seemingly unrelated fields.

The focus of this chapter is how digital tools, technologies, and data can amplify ALIEN thinking. Each component of the framework can be strengthened through a blend of human and digital capabilities to produce outcomes that are far greater than the sum of the inputs.

Capturing new insights, which is at the core of attention, can be augmented by nonjudgmental monitoring technologies. Recall how Nestlé's digital acceleration team uses social media monitoring tools to follow what people are saying about its brands in real time, around the world. Today, the company is in a position to get ahead of unanticipated and potentially damaging events like the public relations attack from Greenpeace. Information is the raw material for ideas. Digital tools excel in focusing the attention of ALIEN thinkers on collecting and presenting those ideas.

The link between digital tools and levitation is less clear. On the one hand, digital tools can be extremely distracting, working against the idea of a time-out. Earlier, we recommended taking a ten-minute walk every day (without your phone) to clear your mind. Digital applications constantly bombard us with information, making it hard to refocus and rise above what we're doing. But it doesn't need to be this way. Some digital tools can oxygenate our minds, helping us disengage from what distracts us. For example, an app that promotes mindfulness can actually facilitate levitation. And while disengaging from your regular activity can be achieved by spending a week in the mountains, it can also be achieved at work or at home by listening to a podcast or taking an audiobook time-out.

Your imagination can be stretched by easy access to powerful technology. Digital tools open up new creative possibilities. For example, to conceive a vetting service for people with criminal histories, Teresa Hodge had to make the connection with technology. Initially, she was simply responding to a one-off vetting request from a community-based bank. Her reaction after spending three days personally evaluating just one person was "This is impossible. We can't do this. It doesn't make sense."[6] The exercise would have remained a one-off and not even have qualified as an insight if she hadn't imagined a tech solution that promised to make such vetting both cheap and scalable—at which point it became a breakthrough idea.

Experimentation can be digitized and accelerated at a fraction of the normal cost. Digital twin technology exemplifies this ability to de-materialize, de-risk, and speed up experiments. NASA saves hundreds of millions of dollars a year by testing spacecraft and other equipment in simulators rather than building and testing physical prototypes. In most cases, digital experimentation is not replacing humans but instead amplifying their ability to set up different conditions, collect data, and interpret results.

The navigation challenges of scaling great ideas and transforming them into breakthrough solutions can be overcome with digital means. Digital tech can be used to help overcome the inevitable (and often underestimated) resistance that ideas face when exposed to the real world. For example, the process of convincing skeptical stakeholders that a breakthrough idea will work can be supported by data from digital feedback sessions or social media campaigns. Dashboards can visualize results, and collaboration tools can help you share supportive data. The most successful ALIEN thinkers today, like Melinda Rolfs, understand the power of digital technologies to inspire, test, improve, and promote their ideas.

THE GOOGLE STORY

To further illustrate the power of human-digital collaborations, let's look at a high-profile success story: Google.

We were teaching an e-commerce class in 1999 when a student shared his enthusiasm for a new search engine. At the time, Google was less than two years old and not very well known. We tried it and found that the student was right. Google was orders of magnitude better than the alternatives. How was this possible?

In the early days of the internet, finding what you were looking for was a major headache. The internet was expanding exponentially and indexing sites like Yahoo! were falling behind, so search results were typically out of date. Most search engines of the time, including Excite, Lycos, and AltaVista, were pretty inaccurate. In the mid-1990s, search engines used the content of a site to assess its relevance to a particular search query. So site owners tried to trick search engines by peppering their pages with sought-after search words.

Larry Page and Sergei Brin, then graduate students at Stanford University, experienced the frustration of poor search results firsthand as heavy users of the internet. Page decided to focus on the internet search problem for his thesis. Rather than join the throng of computer scientists trying to improve search accuracy through existing methods, Page stepped back to view the problem from a new angle.

During this time, Page was switching between two worlds—academia and entrepreneurism. This duality allowed him to reframe the challenges he needed to solve in each domain. Academic responsibilities at Stanford forced him to take time away from his work at Google. Managing a fast-growing start-up meant that he had to step away from his studies for long periods. Just switching between activities is, in itself, a form of levitation because it imposes mental and physical breaks.

In Page's case, switching between activities infused Google with fresh perspectives from his thesis work. In many of his courses, he had to write papers that built on existing academic work. He always found it challenging to know which papers to include in his analysis or which papers were the most insightful on a particular subject. To address this problem, the academic world had developed a practical approach. Papers typically cite dozens of other papers whose work they build on. It was well known that the most frequently cited papers were also the most influential. Hence, the number of citations a paper received was used as a proxy for that paper's importance.

Page's insight was to apply this approach to web searches. In other words, he used his imagination to link two unrelated fields.

Inspired by citation analysis, he theorized that linking patterns on the internet would reveal which sites were most relevant to a given search. The solution, however, was not just about who was linking to whom but—more critically—the prominence of who linked to whom. If a site that links to your site is, in turn, highly linked to from other sites, then it should carry more weight than a site with few links to it.

The idea was a good one, but the mechanics were challenging, which was why Page brought in Brin, a mathematician. Page and Brin worked together to build a ranking system that took into account not only the raw number of links to a site but also the link count of the originating page. Using a combination of mathematics and computer science, they invented the PageRank algorithm (named after Larry Page, not webpages).

Human users of the internet created the links.

Google engineers built the algorithms.

Machines collected and analyzed the data in real time.

PageRank worked well, but it wasn't perfect. Soon, smart web developers were trying to game the system by building sites with many inward and outward links. Google was harder to fool than its contemporaries, but it was not infallible. As Google grew more popular, its engineers continued to tinker with the PageRank algorithm, but website developers were always close behind.

After a while, Google engineers engaged in systematic experimentation to improve the PageRank system and the other algorithms they had developed—like the emerging advertising platform, Adwords. They made extensive use of A/B testing, where one set of users would see one version of the site while another would see a slightly different version. Google could then measure which version was most effective.

The company continued to build its experimentation capability, and by 2008, it was running up to two hundred concurrent experiments on its website.[7] By 2019, it was running millions of controlled experiments a year across its different web properties, most of which were highly or completely automated. Eventually, Google's move into this line of business evolved into Controlled Experiments, an automated A/B testing service that the company began offering to customers in 2012. Controlled Experiments was rebranded as Google Optimize in 2016, and by 2019, it was able to run more than one hundred simultaneous experiments for customers.

COMBINING DIGITAL AND HUMAN CAPABILITIES

Digital tools and technologies were able to amplify the impact of Google's success. A human insight, like applying principles of reverse citation analysis to internet searches, could only become a reality through advanced computational analytics. The Google algorithms could only stay ahead of competitors and manipulative users through digitization and automation. Scale could only be achieved through high processing power, large storage capacity, wide bandwidth, and fast networks.

The Google story also illustrates the importance of combining digital technologies with human insights to amplify the impact of ALIEN thinking. It's hard to imagine that Page and Brin could have achieved their success without advanced digital technologies. However, those technologies, in isolation, would never have made the PageRank breakthrough. Without human ingenuity, we might still be using AltaVista to search the web today.

Google continues to find ways to combine human and machine capabilities to solve large problems. Google Translate, for example, combines artificial intelligence with human verification to improve the quality of machine translation. And reCAPTCHA, acquired by Google in 2009, uses human verification to validate the digitization of millions of print publications, including the complete archives of the *New York Times*.

But if we step outside the realm of the internet giants, are digital tools still as relevant for creating breakthrough solutions?

We believe so.

The ALIEN thinker's digital toolkit consists of three key amplifiers. These amplifiers take the process of creating solutions to new levels of comprehensiveness, speed, and effectiveness.

Digital Amplifier #1: Data Without Observation

The first digital amplifier involves data collection.

Direct observation, the traditional mode of data collection, is a time-consuming business. Anthropologist Margaret Mead spent nine years in Samoa, Indonesia, and New Guinea studying the customs of Indigenous populations. Jane Goodall spent fifty-five years in Tanzania learning about the habits of primates. Thankfully, the digitally savvy ALIEN thinker needn't spend so much time gathering data. Digital tools and technologies provide new ways to interpret behavior without having to directly observe it. Cameras, beacons, sensors, cookies, wearables, and other connected devices allow ALIEN thinkers to collect data remotely with high levels of accuracy. This approach to data collection is certainly helpful for attention, and it can also contribute to the effectiveness of imagination, experimentation, and navigation.

We call this enhanced data-collection ability *digitally enabled awareness*.

Digitally enabled awareness is an advanced sense of what is going on, a keen understanding of the trends shaping the competitive environment, and an ability to see new and emerging opportunities and threats that may not be visible to most people.

Even the most energetic person cannot be everywhere, but digital tools can, and they've made the lives of ALIEN thinkers much easier. Screening software, data-scraping tools, sensors, and other connected devices can help them become much more aware of their environment.

For example, social media tools and applications allow for the observation of a whole new realm of behavior that didn't exist a few years ago. Facebook, Snapchat, Instagram, WeChat, Twitter, Google, and LinkedIn provide fertile ground for ALIEN thinkers to observe and build insights from online social behavior. Indeed, digital tools and technologies can be used to observe behaviors that are unobservable in a traditional sense—such as social media interactions and online shopping.

A story about the Toronto Raptors basketball team illustrates how value can be derived from the insights provided by sensors.[8] The team decided to embed sensors in the shirts of their players during practices. To their surprise, they discovered that the players moved in a forward direction only 15 percent of the time. The other 85 percent of the time, they moved sideways, backward, or diagonally. These non-forward movements required the use of muscles that the team wasn't focusing on in their strength and conditioning programs. The team used this insight to change the players' training habits, devaluing traditional exercises such as wind sprints and encouraging exercises that better mimicked game movements. As a result, injuries dropped the following season. By 2018, the team was using sensors to track players' movement, sleep, and playing style, as well as environmental conditions, to predict the probability of each individual player becoming injured during a season.

Another example of digitally enabled awareness involves the Swiss phone giant Swisscom. Like most telcos around the world, Swisscom was looking at a bleak future.[9] Average revenue per user, a standard measure of telecommunications industry performance, had been dropping steadily. New revenue from online data metering were failing to compensate for falling revenues from traditional sources such as home phone lines, voice calls, SMS, and TV services. The company remained profitable, but at an ever-decreasing rate. Swisscom needed to find new business opportunities.

At the same time, the Swiss town of Montreux was facing an ever-worsening traffic problem. As in many areas of Switzerland, a vibrant economy combined with a growing population and limited space for roads was causing the town's ancient streets to become clogged with traffic during the morning and evening commutes.

The situation was compounded by an alpine problem. Montreux is nestled between the east end of Lake Geneva and the foothills of the Alps. The only highway between the French part of Switzerland, including the airport in Geneva, and the main alpine ski resorts was just above the city, where it cut through the closest mountain. Traffic in this tunnel often ground to a halt during winter weekends as throngs of sports enthusiasts made their way to and from the slopes. When the highway clogged up, many commuters opted to drive through the center of Montreux instead, the only alternative to the tunnel. The problem for the town was that it had no idea which traffic was local and which was passing through.

Traffic is expensive to monitor. It typically requires laying cables across a road to measure the number of vehicles, plus workers to watch the roads and make notes on traffic flow using clickers and other devices. These approaches are fine for measuring vehicle volumes, but they say nothing about the source or destination of the traffic. A couple of failed attempts to address the traffic problem had soured the town on the traditional approaches and vendors. The town was stuck. There had been decades of discussion about building a bridge or underground tunnel to reduce the transit traffic, for an estimated cost of 150 million Swiss francs, but it wasn't clear whether that would solve the problem.

As a resident of the area, Raphael Rollier, then digital innovation and transformation manager at Swisscom, was keenly aware of the town's traffic problems. The roads were simply not designed for the high volumes of traffic that flowed along them. One day, while sitting in traffic, he came up with an ingenious solution. It occurred to him that Swisscom was already collecting location data from all its mobile phone subscribers but was doing little with it. Since Swisscom had a penetration rate of about 60 percent of the Swiss population, this location data could be used to create models of traffic flow through towns like Montreux. Rollier approached the town and offered his company's services. Because it was already collecting the data, Swisscom could provide a traffic analysis for a fraction of the cost—and in a fraction of the time—of traditional traffic analysis.

So Rollier worked with town officials to experiment with different road configurations and traffic signals to reduce congestion during peak periods. Because the traffic was remarkably stable from day to day, the experiments were relatively easy, and monitoring the results was cheap and fast.

Within two weeks, Rollier's team was able to aggregate and analyze a year's worth of traffic data in and out of town by tracking mobile phones. Immediately, certain patterns emerged: except for thirty minutes in the morning and forty-five minutes in the afternoon, Montreux's roads were able to accommodate the traffic volume. In fact, a handful of intersections were responsible for most of the problems. Based on this data, Rollier concluded that a tunnel would not solve the traffic problems and, therefore, would not be worth the investment. Instead, the data suggested that the issues could be improved with a combination of dynamic traffic-light management and better public transportation.

The data further revealed that only 30 percent of traffic was "in transit," meaning vehicles that didn't stop in the town center. And this was only a problem for a few weeks per year, during the height of ski season. As a result of a few weeks of experimentation, Montreux eventually implemented a new traffic system and improved its public transportation options, which dramatically reduced the congestion issues.

Rollier's imaginative insight was the realization that Swisscom could use information collected in its primary business to solve a problem and provide a service for a completely different application. (Swisscom is currently rolling out similar services across Switzerland.) As Rollier noted, there are significant benefits to blending digital and human problem-solving approaches. "I see urban planning of the future as a combination of digital and traditional approaches. Digital tools can enrich human approaches rather than replace them."[10]

Swisscom is now combining its mobile phone data with data from sensors embedded in roads, cameras, and other technology, and from human observation. Why not use only the mobile data? As it turns out, phone signals have a hard time differentiating between cars, bicycles, buses, and pedestrians. Also, the level of precision that can be achieved from mobile phone signals is fine for a relatively large municipality like Montreux, but it's not good enough for more compact areas.

We regard direct observation as critically important, but it can be augmented by digital tools, connected devices, sensors, and the like to gain a more comprehensive understanding of actual behavior. These devices can provide the ALIEN thinker with a whole new set of data on which to build or test insights, whether for supporting attention, levitation, imagination, experimentation, or navigation.

Digital Amplifier #2: Insights Without Bias

There's a famous experiment designed by the behavioral economists Daniel Kahneman and Amos Tversky that we like to re-create in our classes. In the experiment, they describe a hypothetical person named Steve: "Steve is very shy and withdrawn. He is invariably helpful, but with little interest in people or in the world of reality. Steve is a humble and tidy soul. He has a need for order and structure, and a passion for detail."[11]

We ask our students whether they think Steve is more likely to be a farmer or a librarian. This question typically splits a room. About half feel that Steve is more likely to be a farmer, while the other half chooses librarian. There are good explanations for both perspectives, supported by common stereotypes: that librarians tend to be shy and feel comfortable within the structured environment of a library, and that farmers are solitary folk who are more interested in their crops and animals than in other people. Convincing arguments can be made for either side.

Very few people, however, consider the underlying population data. In virtually all places on earth, farmers vastly outnumber librarians. Moreover, farmers are more likely to be male and librarians female. Thus, if you consult the data, Steve is *much* more likely to be a farmer than a librarian.

Most people get stuck on identifying clues from the description, supported by stereotypes they have about farmers and librarians. (Kahneman and Tversky called this the "base-rate fallacy.") As humans, we are hardwired to overestimate the importance of data that's in front of us and underestimate the importance of data that's not. It turns out that the base-rate fallacy is one of hundreds of biases that humans have developed over millennia.

Many of these biases evolved to help us in stressful or dangerous situations. Prioritizing the information in front of us is very helpful when we're in the presence of a hungry predator. It's less helpful, however, when we're seeking new insights. Fortunately, digital tools can help us to recognize and avoid these biases.

Let's consider an example of how analytics can help ALIEN thinkers with the process of sensemaking for a very important decision: finding a life partner.

The traditional process for selecting a partner is relatively random, intuitive, and emotional. Online dating has changed this dynamic, and now there is much more science behind matchmaking. The results have been dramatic: in 2017, almost 20 percent of marriages in North America originated from online dating, up from 5 percent in 2015.[12] In addition, online dating sites provide a treasure trove of demographic and behavioral data that can now be used to assess the likelihood of a successful match. For the first time in history, we are able to scientifically break down the emotional process of finding a partner.

The Harvard data scientists behind the dating site OkCupid analyzed responses to questions from more than a million members and identified two questions that strongly indicate whether a match is likely to succeed—when two parties give the same answer, the chances of a successful relationship vastly increase.[13]

Can you guess what the questions are?

Perhaps the following questions popped into your mind: *Do you want to have a family? How important is your career to you? Are you a morning person? How often do you exercise?* In reality, these questions were very low on the list of predictors of lasting relationships. As it turned out, the two most important questions were (1) *Do you like horror movies?* and (2) *Have you ever traveled around another country alone?*

When you think about these questions, you may come up with some reasons why they predict successful relationships. Perhaps horror movies are a proxy for excitement without risk. Maybe traveling alone is a proxy for independence. The reasons, however, are less important than the fact that the questions work. If you and a prospective partner agree on these two questions, you are more likely to stay together.

Finding nonintuitive insights that cut through the limitations of human bias is a key benefit of digital tools. Big data and analytics systems are being built for the express purpose of unveiling surprising, nonobvious, and creative patterns and linkages within data. Studies have shown that humans struggle to accurately predict which couples will stay together because they lack the analytical abilities to cut through massive quantities of data.[14] Fortunately, today's ALIEN thinkers have many more tools in their arsenal than their analog predecessors.

Digital Amplifier #3: Scale Without Compromise

Even when insights are uncovered, they are often seen as interesting anomalies. Tangible benefits come with scale. Digital tools can help transform an interesting but small insight into something large and impactful.

The investment industry has been a trailblazer in this area. Many new analytics tools have been developed to search vast quantities of financial data for nonintuitive patterns that might uncover overlooked investment opportunities.

A leader in this area is BlackRock, the world's largest asset manager, with over $6 trillion in assets under management and investment positions in thirty countries.[15] The key architect of the analytics capability is the firm's chief operating officer, Rob Goldstein. Goldstein oversaw the development of BlackRock's secret investment weapon, Aladdin (the name is an acronym for "asset liability and debt and derivative investment network"). Aladdin sucks in all kinds of quantitative and qualitative data, builds risk-adjusted real-time investment strategies, and provides insights down to a single trade in a split second.

Regarding the ability to scale, Goldstein said, "I believe the new technologies that have emerged over the past few years . . . have really unlocked the opportunity to evaluate and manage investments at scale in a whole new way. It's just a different scale factor."[16]

Goldstein's team developed digital tools to uncover nonintuitive patterns among global equities. This allowed them to leverage these patterns to make smart trades before the wider market caught on. The following quote comes from a BlackRock report from October 2015:

One such strategy that has proved effective begins with identifying clusters of seemingly unrelated stocks that share common economic return drivers. . . . To find fundamentally related companies that are not obviously correlated, we employ text-mining algorithms that can interpret vast quantities of written materials, such as company reports, regulatory filings, blogs and social media. These tools enable us to identify securities that are exposed to similar return drivers despite differences in their industry classification, country of domicile, market capitalization and position in the supply chain, among other factors. Our analysis has uncovered non-intuitive relationships between seemingly unrelated securities, such as a French testing and inspection company, a Dutch maritime infrastructure corporation, a Silicon Valley tech firm, and a U.S.-based global financial services company. While this may seem like a disparate group, it turns out that the testing company's clients span sectors from industrials to banking and that part of the Silicon Valley firm's business model includes offering similar testing services. Only a deep analysis of their businesses reveals the links that tie these firms across multiple industries.[17]

Sound familiar?

It should. It's similar to how we've described ALIEN thinking throughout this book—a process of careful observation across a variety of information sources, a search for counterintuitive insights, a rejection of conventional answers, and a careful testing of hypotheses. These steps are exactly what Aladdin has been designed to follow, but on a massive scale.

It's interesting to note that Aladdin analyzes not only transaction data, such as trade volumes and prices, but also unstructured data, such as company reports, blogs, and social media posts. Analysis at scale has traditionally been a very human-centric activity because computers were not able to easily capture or analyze unstructured data. Nonintuitive inferences were beyond them because they were simply too logical.

This is no longer the case. Today's analytical systems can take a variety of information as inputs—from emails and websites to audio and video—all in real time. The inferential tools and algorithms that underlie the ability to make connections, recognize patterns, and uncover latent insights have also improved enormously. They are, in effect, industrializing the process of attention, levitation, imagination, experimentation, and navigation.

WINNING WARS WITH COMMUNICATIONS AT SCALE

Stanley McChrystal doesn't come across as your typical ALIEN thinker. He's a clean-cut former military man who spent his whole career in the US Armed Forces before retiring as a four-star general in 2010. Yet he potentially did more to shake up the way the armed forces operate than anyone else in recent history.

Let's pick up the story in 2003, when he was asked to lead the Joint Special Operations Command (JSOC), a task force that oversees several elite counterterrorism units, including the Army Rangers and Navy SEALs. JSOC was leading the charge to counter new threats posed by global terrorist organizations, particularly Al Qaeda in Iraq (AQI).

McChrystal quickly realized that AQI was winning. Despite the fact that the US Armed Forces had the world's biggest military budget, the most sophisticated equipment, the newest technology, and the best-trained soldiers, they were losing the battle on the ground to more agile forces. "We were the most elite force in the world, with unrivaled discipline, training, and resources, but we were facing an organization that was not our mirror image," said McChrystal. "AQI was a completely different beast—a dispersed and nimble network. In order to win, we needed to go after the whole network, not only the top few leaders."[18]

The US forces were taking a traditional approach to fighting AQI—doing the same things they had always done, but better. However, it wasn't working. They were constantly being surprised by attacks. By the time the intelligence would arrive, it would be too late. When they finally got boots on the ground, the insurgents were long gone. The forces' traditional advantages in size and firepower were of little use against this more agile enemy.

McChrystal acknowledged that they needed to do things differently. But how to get better when they already had the best of everything?

The main issue, he realized, was a failure of communication. Silos in the armed forces meant that information was not being shared between functions or forces. The units in the field did not interact with the intelligence analysts or the big-picture decision-makers in Washington, DC, or at JSOC headquarters in Fort Bragg, North Carolina. The air force, navy, army, and marines harbored deep reservations toward each other. A culture of mistrust, combined with a strong focus on secrecy and a history of need-to-know communication lines, meant that little information was being shared, and when it was, sharing happened slowly.

To improve the communication within the task force, they "redesigned [the] headquarters, opening up [the] physical space to include everyone, and mounted state-of-the-art technology to display information and allow for video-conferencing across multiple screens," said McChrystal. "We had an almost fanatical focus on sharing information across the Task Force, from the analysts in DC to the operators on the ground in Mosul—we held an organization-wide 90-minute meeting every single day, sharing updates and lessons learned from around the globe with over 7,000 people."[19]

These meetings were a radical change from the norm. Meetings usually happened on an ad hoc basis, with a small number of people. Instead, McChrystal included everyone involved directly in the war in Iraq. Many people thought this was crazy. They questioned the potential for information leaks. They complained about the time commitment. At the beginning, few people participated, so McChrystal had to stress the importance of the meetings in the most noticeable way: by always personally attending them and never postponing or rescheduling. After a short while, attendees numbered in the thousands.

Soon the increased flow of information led to better and faster decisions. Often, when a problem came up on a call, someone had a solution. People became more confident in speaking up and sharing what they knew. Trust between the forces and functions began to blossom. Most important, the war started to go better, and US forces were able to get close to matching the pace of AQI and other terrorist organizations.

McChrystal had to apply unconventional thinking to revise the approach to communications with JSOC. He needed to pay attention to the problem by zooming in and out of the day-to-day battles. He needed to remove himself mentally from a whole career of doing things a particular way and use great creativity to identify new solutions. He needed to experiment with different approaches, technologies, and formats before he found the ultimate solution. Most important, he needed to convince an influential group of skeptical senior military leaders to change their approach.

None of this would have been possible without the digital communication technologies that supported his strategy. Imagine the technical requirements of running a global, live ninety-minute meeting with up to seven thousand participants each and every day. Imagine the security protocols that needed to be in place. Imagine the bandwidth requirements in a variety of locations, including very remote field settings. All of these had to be developed to transform a breakthrough idea into a workable solution.

The solution that McChrystal invented was inherently human—a live conversation among colleagues. But this solution was facilitated by an advanced set of technologies, many of which had already existed, but some of which had to be built.

THE ROAD AHEAD

IBM's Watson can outperform physicians in diagnosing certain cancers. Google can finish your sentence for you. Siri can guess your mood. Facebook can describe your personality. Database environments, like Apache Hadoop, have been developed to capture all forms of structured and unstructured data across multiple locations. This data is made accessible to organizations with tools such as MapR,

Cloudera, and Hortonworks. Analytics engines sit on top of these vast data sets and mine them for valuable insights.

In some respects, digitization has industrialized the insight development process.

ALIEN thinking is a major adjustment for most managers. It requires them to challenge their most deep-seated assumptions about how to manage and make decisions. Fortunately, a set of digital tools and technologies is available to support them today. These tools help ALIEN thinkers navigate the route from interesting insights to impactful solutions. Ideas are interesting, but value can be captured only when those ideas are brought to life and implemented at scale. Digital tools and technologies can assist in turning insights into action via three digital amplifiers: data without observation, insights without bias, and scale without compromise.

⊷⊶

KEY TAKEAWAYS

- ALIEN thinking can be amplified through the combination of human and digital capabilities. This amplification can happen through three mechanisms.

  » Digital amplifier #1: data without observation. Digital tools can enable the remote observation of behavior and collection of data. We refer to this as *digitally enabled awareness*. This awareness can be facilitated by sensors, beacons, and other connected devices, or via observation of behavior on social media and other collaborative applications.

  » Digital amplifier #2: insights without bias. Human behavior is rife with biases and inconsistencies. Digital tools, while not immune to preprogrammed biases, tend to be less prone to systematic errors.

  » Digital amplifier #3: scale without compromise. Ideas are interesting, but value can be captured only when those ideas are brought to life and implemented at scale. Unfortunately, moving promising ideas from concept to scaled implementation is a constant challenge. The platform characteristics of many digital tools allow them to scale quickly and cheaply.

- As the world becomes more fast-paced, unpredictable, and complex, ALIEN thinkers need to consider how new generations of digital tools and technologies can amplify the innovation process.

QUESTIONS TO ASK YOURSELF

Digital Amplifier #1: Data Without Observation

1. Do I have an opportunity to collect data automatically—for example, from physical or digital sensors?
2. What data is available that I am not currently collecting?
3. Can I collect objective data on how customers/employees/ stakeholders behave?
4. Am I overreliant on my own observations?

Digital Amplifier #2: Insights Without Bias

1. How are my biases affecting my judgment?
2. Can I use analytics to support my intuition?
3. Can I use digital tools—for example, figures, graphs, or dashboards—to visualize the data?

Digital Amplifier #3: Scale Without Compromise

1. Can I use digital technologies to expand the scope of my ideas?
2. Can I scale my ideas faster using digital technologies?

TEN

Greet Your Inner ALIEN

WHEN BERTRAND PICCARD APPROACHED AVIATION COMPANIES about building a solar plane that could fly through the night, he was told it was impossible. It took their experts about five minutes to calculate that you couldn't capture and store enough solar energy to keep a plane airborne for twenty-four hours.

As he struggled to pursue his dream, their voices sometimes echoed in his mind. Looking back on his achievement, Piccard acknowledged, "Honestly, it was *very* difficult. I never imagined it would be so difficult, it would be so expensive, and it would be so long. There were so many setbacks, and so many moments where I thought, 'We're not going to do it. It'll be a miracle if it works.' The plane was so fragile. When there was too much wind on the ground, nobody could hold the plane and it was in danger of being destroyed. And there were so many moments where we really were very close to failure."[1]

In addition to all the external voices telling you it can't be done, you also have internal voices to contend with. These can make you doubt the value or feasibility of your endeavor, or question your ability to see it through. Because of how the mind works, it's not easy to be an ALIEN thinker.

THE INTERNAL BARRIERS

We've already talked about the blinders or biases that make it difficult to think unconventionally. But there are other mental barriers to overcome, rooted in your specific psychological makeup: your vulnerability to negative emotions and traps relating to your supposed strengths.

The challenge of managing yourself, often discussed in writing about leadership, is conspicuously absent from the literature on innovation and entrepreneurship.

Delivering breakthrough solutions is not just an innovation journey; it's also an inner journey. Along the way, even ALIEN thinkers will be beset by fears, doubts, regrets, and frustrations, or, conversely, by misplaced feelings of confidence, certainty, and invulnerability. Your emotions can cause problems, and so can your personality traits. Paradoxically, you can run into trouble because of your strengths as well as your shortcomings.

To avoid getting derailed, you must learn to manage your emotions and yourself.

THE FEAR FACTOR

As an innovator, most of the mental and emotional stress you experience is rooted in fear. Fear is a critical emotion that keeps you safe and comfortable. It is also a powerful deterrent to unorthodox action.

When setting off on a journey of innovation or discovery, you will have to confront your fears about what might happen. An extreme example is Dr. Billy Fischer, who was asked by the WHO to go to Guinea to inject new thinking into managing the deadly Ebola virus epidemic. Fischer had experience in resource-constrained environments with critically ill patients, but he was a respiratory specialist, and Ebola is not a respiratory virus.

"That was a hard conversation with my wife," he recalled.

She said, "Why you? You're the wrong person to be sent to this place." She was right: I had very little viral hemorrhagic experience. But I had the skill set that was going to get me into the position where I might be able to do some good.

I told my wife, "Look, I'm *exactly* the right person to do this, because I know how to take care of really sick people, I know how to keep myself safe, and I've worked in these conditions before." She was not sold.[2]

It took about a week to sort out the vaccinations and visa—and that week proved emotionally grueling. Fischer remembered, "That week before I got there . . . was the most intense experience of my life. I actually told the guy at WHO, 'You need to just get me on a plane, otherwise, I'm not going to go. If I think about this any longer, I'm going to get off the list . . .' I honestly think that if I would have waited one more day, I wouldn't have gone. The fear was overwhelming."[3]

Bertrand Piccard was also putting his life at risk when he chose to solo pilot a solar plane through the night. Of course, the stakes don't have to be life-or-death to trigger anxiety. Many of the ALIEN thinkers featured in this book took potentially life-changing risks: Narayana Peesapaty gave up his safe career as a researcher and mortgaged his house in order to manufacture edible spoons; Terra-Cycle's Tom Szaky dropped out of Princeton University and poured all his savings into a "worm gin" contraption to produce his worm poop organic fertilizer. Malcom McLean walked away from his successful trucking company to build a new business based on some untested ideas about shipping. KidZania cofounder Xavier López Ancona abandoned his successful career at GE Capital, and Wysa cofounders Ramakant Vempati and Jo Aggarwal gave up well-paid jobs at Goldman Sachs and Pearson Learning, respectively, to create a mental health chatbot.

Their sacrifices differed in size and reversibility, but they all involved overcoming fear. Although assessing the merits of your idea and its potential may be difficult, you generally know what you are forsaking to pursue it. Safety and comfort are usually the first casualties. And your brain is hardwired to take such risks very seriously.

FUTURE REGRETS

Regret is an uncomfortable and memorable emotion. The experience of regret is linked to feelings of self-recrimination: *I should have known better. I brought it on myself. If only I could start over. It won't happen again* . . .

When your decisions turn out badly, you regret your wasted effort, your mistakes, or your bad luck. The desire to avoid painful feelings of regret often encourages future caution and conservatism. Recall the lesson learned by the Hoover VP for Europe who lamented turning down the chance to acquire James Dyson's bagless vacuum cleaner: "I do regret not taking the product technology off Dyson. It would have lain on the shelf and not been used."[4] Even with full knowledge of the spectacular outcome, he regretted not making a decision that would have maintained the status quo.

Regret is not just a painful emotion you feel when looking back. Psychologists have found that you also project yourself into the future and experience "anticipatory regret." You imagine the regret that you will feel if a decision proves to be a mistake.[5]

The fear of future regret looms large in many decisions and nondecisions. Studies show that you anticipate far more regret than you actually experience.[6] And the expectation of future regret is all the stronger when you make an unconventional choice. It feels harder to escape self-blame if you make a choice that others would not have made.

One executive expressed the dilemma this way: "I am on an escalator going up, and in order to innovate, experiment, or learn from failure, I have to get off it. And I don't know if I can actually get back on afterwards. . . . And I will kick myself for jumping off."[7]

You punish yourself in advance. When you have an opportunity to deviate from the conventional path, your brain responds with anxiety and doubt to dissuade you from taking unnecessary risks. You are on a predictable professional trajectory, with a settled family life and financial security. You know what you're giving up. You're not sure what you may be getting.

Typically, anticipatory regret keeps you thinking along conventional lines. But that cognitive distortion can be turned to your advantage.

Instead of evoking the anticipatory regret that comes with envisioning failure, you can reframe the anticipatory regret: imagine how you'll feel if you don't even try. Imagine looking back at this missed opportunity to make a difference. Imagine that you didn't follow your heart because you decided to play it safe. Scholars call this *existential regret*.[8]

Consider Narayana Peesapaty in India. The rural-development researcher tried to combat groundwater depletion in his region by reviving demand for millet at the expense of water-guzzling rice. His idea of millet-based organic cutlery took a decade to turn into a viable business.

In retrospect, he deeply regretted the sacrifices he imposed on his family, especially his daughter, so that he could follow his passion. He remarked, "My only daughter has forgotten her demands and simple needs she would have been showered by me otherwise. I know I cannot give her back her lost childhood or pamper her like teenagers usually are by their fathers."[9]

What carried him through this painful period during which he was building his business was the regret he would have felt if he had done nothing—groundwater levels would have been halved within three decades, degrading the quality of the soil and the river water in the process. "By the time my daughter is my age, the planet will become uninhabitable. My fight is to give their generation a fighting chance."[10]

For Peesapaty, the prospect of *future* regret was so powerful that it enabled him to overcome the *experienced* regret of inflicting hardships on his daughter. What kept him going was the bigger goal: the cause.

WHERE ALIENS FEAR TO TREAD

The problem with fear is that it's overprotective. It evolved to deter you from risky decisions in survival situations. Today, many of those primal threats no longer exist. (When was the last time you were stalked by a saber-toothed tiger?) So instead, you focus on second-order risks to your comfort, status, and self-esteem. In addition, you typically overestimate the consequences of failure and underestimate your capacity to quickly recover from setbacks.

When pondering a new venture, there are two broad sources of fear: (1) doubts about the idea itself, and (2) doubts regarding your ability to bring the idea to life.

Uncertainties about the validity and potential of an idea are inevitable. For example, although Chris Sheldrick was utterly convinced about the superiority of his three-word address system (What3words) compared to existing alternatives, he had doubts about whether it would be widely accepted. When he pitched the idea to investors, a common reaction was, "Well, this is slightly delusional, and you guys must be mad if you think this is going to get adopted."[11] Sheldrick's challenge was to convince people not only to buy a new product or service but to adopt a new standard that was a radical departure from the global address system and GPS coordinates.

Instead of worrying about developing an ecosystem, he focused on finding partners for whom the existing alternatives were a clear pain point. Early adopters were a motley bunch that included the UN (for disaster relief work), the Mongolian government (for its postal services), and Mercedes (for its voice-recognition vehicle navigation systems). These use cases validated the system, but it still needed to build critical mass. As Sheldrick conceded, "Even for people who love What-3words and know their three-word address, to actually get them to be regular users is a huge behavioral change. You have to get into your car and it needs to be normal to say 'Navigate me to table.cabbage .spoon.' That kind of feels a bit bizarre to us."[12] His solution is to take things one step at a time and view this as a normalizing process. "I think we are just moving people along the conveyor belt from 'crazy, slightly wacky, but interesting and cool' to 'really useful, can use it in lots of places, and I actually think that this is going to be a system.'"[13]

Compounding doubts about the idea are anxieties about your personal ability to develop a successful venture. Do you have what it takes to make it happen? Even seasoned entrepreneurs can doubt their capacity to run a new venture. For example, Teresa Hodge had previously worked as an HR professional and as an entrepreneur. But that was before spending five years in prison. "I had envisioned by the time I was in my fifties, I would be slowing down . . . instead of starting over."[14] She worried that she might not have the energy needed. Also, she was planning to target the civic technology space, yet incarceration had seriously eroded her digital skills. "I knew I was going to have to come home and integrate technology into my life . . . in order to play catch up," she said.[15] What helped her overcome her reservations was focusing on her strengths. "I'm not a data scientist, I'm not a researcher, I don't write code," she conceded. "But I have enough experience in business and I have the firsthand experience of being incarcerated in this country. So I figured it out."[16]

Beyond capability concerns, you may also worry about your perceived credibility with investors or partners. You may fear the judgment of others and doubt that you will be taken seriously. Recall the case of former car mechanic Jorge Odón. He cringed at the prospect of sharing with obstetricians his far-fetched alternative to forceps to help extract babies from the birth canal. Odón later remembered his misgivings: "You feel a bit crazy when you invent something. You tell yourself, 'But if this was on the internet . . . I'm not a doctor. What the hell am I doing here?' You are full of doubts. Doubts prey on your mind and you feel awfully distressed."[17] Would they even listen to him? And if they did, would they perhaps try to steal his idea? His remedy was to take an engineer friend with him to the meeting.

Such angst is understandable for innovators and disrupters, who are often operating outside their comfort zones—acquiring new knowledge and competencies, leaving behind old ways of thinking, and taking leaps of faith. You are bound to have doubts about whether you can manage an unconventional challenge, as well as anxieties about appearing foolish or inept. As Richard Branson pointed out, "It is normal for entrepreneurs to feel out of control and worry that the rug could be pulled from underneath them at any moment."[18]

MOVING THROUGH FEAR

Fear is a common response when you are on the brink of growth or positive change. Unmanaged, fear can overpower original thinking. It can send you spiraling into a state of worry, self-consciousness, and uncertainty. But fear doesn't have to be a negative emotion or inhibiting force. Fear can also be a driver. Branson highlighted its positive potential: "A touch of the jitters sharpens the mind, gets the adrenaline flowing and helps you focus. It is important not to fear fear, but to harness it—use it as fuel to take your business to the next level. After all, fear is energy."[19]

Fear of failure pushes you to figure out the flaws in your idea. It encourages you to look for people who can provide constructive criticism. The problem is, fear is often experienced as a diffuse and overpowering feeling rather than a specific thought. To leverage fear, you first need to trace its origin.

ALIEN thinkers do not suppress or ignore fear. On the contrary, they pay close attention to it. We've made a broad distinction between fears relating to your idea versus fears about your capacity to enact it. We've also highlighted the fear of missing out by straying from the conventional path and failing. Empirical research with entrepreneurs has refined these three categories and identified seven common sources of fear: financial security, ability to fund the venture, personal ability/self-esteem, potential of the idea, threats to social esteem, the venture's ability to execute, and opportunity costs.[20]

Pinpointing the source of the fear is critical to addressing it. Only then can you start taking anxiety-reducing actions. As Bertrand Piccard pointed out, "Doubts are *so* fascinating. If you have a doubt, it means that you are going to look twice, or twenty times, for better solutions. If you have a conviction and you know it's like this, and you do it, usually it's a straight line going straight to an obstacle—and you fail."[21] If you establish that the fear stems from your individual shortcomings, then learning, information-seeking, or partnering can help to mitigate those doubts by increasing your capabilities.

Becoming an ALIEN thinker means altering your relationship to fear, learning to read it and channel it into action. Inspired by the concentration camp reflections of Austrian psychologist Viktor Frankl, Stephen Covey, author of *The 7 Habits of Highly Effective People*, once wrote, "Between stimulus and response there is a space. In that space is our power to choose our response. In our response lies our growth and our freedom."[22]

In order to be an ALIEN thinker you have to learn to manage your emotional response, not just to what *might* happen, but also to what *has* happened.

THE OTHER F-WORD

Besides fear, the other big emotional derailer is frustration. It is the result of disappointments and regrets about what has happened or is not happening. It's a feeling of not progressing or of even going backward—of having made the wrong choices.

Bart Weetjens recalled the frustration he experienced as he struggled to secure funding for his brainchild: "When I first approached people with the idea of using rats to find land mines they laughed at me. I spoke to the military in Belgium, where I'm from, and several other organizations but they said I was crazy. . . . I was frustrated when my idea kept getting rejected. Then, after a few years, I got lucky. I met a former university lecturer and he thought the idea was ingenious. He put me in touch with the Belgium Development Corporation and I was finally given funding to start my research."[23]

Not only did Weetjens have to withstand laughter and rejections; he then had to work with some of the organizations that had previously turned him down. As a practitioner of Zen Buddhism, Weetjens used his training to work through those frustrations and resentments.

Success does not necessarily diminish feelings of frustration. Six years into his journey, Pavegen's Laurence Kemball-Cook noted, "I feel so low sometimes. . . . I feel probably worse than when I first started the project. I'm never happy with it because our technology is not in every street in the world. Until that's happened, I'm restless, I haven't succeeded. That is my goal. But it won't happen overnight, and maybe that's why I get frustrated."[24]

For Kemball-Cook, training for a triathlon proved to be a good way of working off frustration and putting his problems in perspective. "Sport is a great way to separate yourself from those challenges," he said. "If you spend eight hours on a bicycle, cycling up a mountain, it's amazing how many of those problems you've thought about, you've processed, and you've got a solution. If you just run from A to B to C, you forget about that big picture."[25]

Besides stepping back from the disappointments, the other key to fending off frustration lies in your attitude toward failure. When things don't work out as you hoped, it's easy to succumb to negativity. As TerraCycle founder Tom Szaky, observed: "When someone says no to you, the emotional reaction is to be upset and be annoyed. What will get you the most value out of something like that is to ask why. Why did this happen? How could it have been better?"[26]

Resist the urge to move on and put the failure behind you. You have to do a postmortem. What went wrong and why? What assumptions proved false? You may realize that the weaknesses overshadowed the strengths, but that doesn't invalidate the strengths. Maybe you can find new ways of leveraging the strengths.

This forensic mindset helps you process disappointment and bounce back from it. It develops mental fortitude—what Angela Duckworth calls "grit."

Recall the case of Van Phillips, who literally had to pick himself up hundreds of times as he experimented with different designs for his prosthetic cheetah leg. He tried over three hundred prototypes of his prosthetic foot at a rate of about one a week. "Every time a prototype breaks, it's heartbreaking," Phillips admitted.[27] "I would be in a very deep depression, because I had experienced running, and then it would break, and I'd have to go back and start over."[28] You have to look at the failure honestly. After each failure, Phillips would change the design or vary the materials, and try again.

What kept him going was a strong learning orientation and the memory of the fleeting exhilaration of running again. Nine times out of ten, said Phillips, ideas don't work: "It's the dedication to the dream that [you] can do something that will actually pull [you] through to the very end."[29]

Besides falling afoul of your fears and negative emotions, you can also be undone by what you regard as your strengths. They can trip you up or even take you out of the game.

KILLER QUALITIES

ALIEN thinking benefits from personal characteristics that you can develop over time, such as empathy, curiosity, openness, persistence, and persuasiveness. But taken to extremes, these same qualities can become liabilities.[30] You can have too much of a good thing.

There is a dark side to many of the qualities associated with ALIEN thinking. You must be mindful of it.

Attention calls for empathy—the ability to put yourself in the shoes of others. But too much empathy is debilitating. Recall the case of Bart Weetjens. He went to Africa to better understand the situation of local farmers who were unable to access their land because of land mines. He bonded with them and was determined to find ways to help them solve their problems by themselves.

Looking back, he noted the dangers of becoming *too* empathic, particularly for social entrepreneurs. "Entrepreneurs should take care not to be eaten through empathic resonance with the problems they are tackling," he warned. "So in the first place, they must take care of themselves, their own personal sustainability. In the second place, those who are dear to them. And only in the third place, when all these are in balance and well sorted, can they be productive in society in a sustainable way."[31] The immersive nature of innovation can lead you to neglect your personal well-being and the close relationships that support you through the journey.

Levitation depends on intellectual curiosity: the appetite to self-question and make sense of a situation. Recall the avant-garde chef Ferran Adrià, who closed his El Bulli restaurant for six months each year "to reinvent ourselves."[32] In 2011, he closed it for good to transform it into a research lab and exhibition space—not just for cooking, but for the innovation process as a whole.

The center was supposed to open within three years. As of this writing, it's nearly seven years overdue. During those ten years, the self-questioning has not stopped. Adrià launched competitions among top business schools to generate ideas for the future of his foundation. He also organized a series of six exhibitions across the world, telling reporters, "Our project has been reinvented on many occasions thanks to these experiences."[33] As *FT Magazine* once pointed out: "The scope of the El Bulli Foundation project is vast and has been in flux since the restaurant closed."[34]

Taken too far, intellectual curiosity can lead to overthinking. The self-questioning becomes debilitating. You don't settle on a course of action. The cycle of reflection that served Adrià so well in one situation has arguably impaired his progress on a less structured challenge.

Imagination relies on openness to novelty. But you can be *too* open. This criticism has been leveled at Google cofounder Sergey Brin, who has a reputation for "project attention deficit disorder," becoming obsessed with one initiative and then jumping to the next.[35] This tendency is said to have contributed to the Google Glass debacle when Brin got personally involved in running the project, because he lacked the patience needed to hold off on launching a product that was nowhere near ready for prime time.[36]

Another problem with being too open to novelty is overreaching, as illustrated by Demis Hassabis. Before launching DeepMind (now owned by Google), he cocreated the hit video game *Theme Park* and then founded his own gaming company, Elixir Studios. Looking back, he drew a key lesson from that experience. "What I tried to do with Elixir is innovate on all dimensions simultaneously. . . . I wanted to create new graphics engines, new AI engines, make an artistic statement about the game."[37] The product he eventually hatched represented only a fraction of the original vision and took so long to develop that it didn't recoup its costs, despite making money for the publishers. Hassabis conceded, "I was being too idealistic. . . . We bit off too much. You have to pick what dimension of innovation you are going to push hard."[38]

Experimentation demands persistence—the focus and self-discipline to keep testing and adapting different solutions. "When you feel like giving up is precisely the point when everybody else gives up. So it's at that point that you must put in extra effort," said serial inventor James Dyson, who created over five thousand prototypes of his eponymous vacuum cleaner. "You succeed by understanding that other people are also feeling tired, so when you feel tired, you should accelerate. That's when you start winning."[39]

Many of the disrupters featured in this book—notably Piccard, Kemball-Cook, Phillips, Weetjens, and Peesapaty—continued experimenting well past the point where reasonable people would have given up. They succeeded despite warnings from experts that they were wasting their time. This healthy disregard for advice can easily be carried too far. Single-mindedness becomes stubbornness. Selective hearing turns into deafness. DeepMind's Hassabis captured the predicament: "You've got to go through many pain barriers to get anywhere useful. But how do you know when you're running full speed into a dead end? And when to cut your losses and use your learning for something else?"[40]

The navigation phase demands persuasiveness. Disrupters who are naturally assertive, energetic, and enthusiastic find it easier to mobilize partners, investors, teams, and staff. But you can be too persuasive—to the point that you discourage challenges, and not just from subordinates. Think back to the charismatic Dean Kamen and the Segway. He was so persuasive in selling his vision to investors that his company secured enough money to work in secret. To raise fresh funding, there was no need to expose the work to external scrutiny or objections. As a result, there was insufficient pushback on basic questions about its design and use.

The same goes for Elizabeth Holmes at Theranos. Often likened to Steve Jobs, she convinced many high-profile figures to join her board and managed to raise millions from supposedly savvy venture capitalists. Her dazzling boasts about revolutionizing blood testing went unchallenged by board members and investors. They never de-manded an answer to the basic question: Does this thing really work? It didn't.[41]

You can shut off internal dialogue and exploration with your bril-liance, passion, and self-confidence. "Weirdly, you can actually over-inspire people—to the point where they're not thinking rationally about what they could achieve in a certain amount of time," said DeepMind's Hassabis.[42]

Looking at some of the qualities that support ALIEN thinking, you must also realize their limits. You have to be careful about overplaying your strengths. Pushed too far, any quality is a potential derailer. More-over, strength in one domain may be weakness in another. For example, the novelty-seeking tendencies associated with imagination coexist un-easily with the focus and discipline associated with experimentation. The self-questioning and reflective qualities needed for levitation are at odds with the persuasive powers that support navigation.

One way of addressing these tensions is to keep switching between the five dimensions to try to achieve a dynamic balance among them.

It's very difficult to be outstanding across all five dimensions. Strength in one area cannot make up for lack of strength in another. ALIEN thinking requires you to be able to do all of them—though not necessarily on your own.

MANAGING YOURSELF

We each have a blend of strengths and weaknesses, but research on leaders strongly suggests that what turns both into derailers is lack of self-awareness.[43] The same holds true for ALIEN thinkers.

Self-awareness is about understanding your abilities, emotions, and drives. What are your outlier tendencies—those ways of being or thinking that you perhaps take to be the norm but that are actually the exception? And what are your vulnerabilities?

Without self-awareness, you will find it hard to develop the coping strategies that enable you to deliver breakthrough solutions. Do you know how to maximize what you have and acquire what you don't?

Disclosing your shortcomings and idiosyncrasies is one way to manage them. It makes it easier for people to read you and call you out when you slip into a dysfunctional pattern of behavior.

Another common antidote is to surround yourself with individuals who can act as counterweights. If you tend to be disorganized, find someone who is methodical; if you tend to look at the big picture, find a pragmatist; if you tend to be impulsive, find someone more risk-averse; if you tend to be too trusting or open, find someone more skeptical or politically astute.

For example, TerraCycle's Tom Szaky knows he is good at convincing and exciting followers with his passion. He also recognizes the dangers of that ability: "It engulfs people. . . . And when those people believe, they may believe with sort of blinders on. And then you don't get challenged."[44]

To combat this risk, Szaky goes out of his way to encourage criticism. "I have one person in my company, we call him our chief administrative officer. And he and I clash all the time. I mean, it's epic. Not fights per se, but philosophical clashes. And the company is better for it every time, because he forces me to challenge baseline assumptions that I'm not necessarily ready to challenge, because I just want to move on and move on, and I sort of believe my ideas are right."[45] To embolden others to speak up, the discussions with the chief administrative officer are conducted openly. "By having our battles [in] public, it shows people that it's totally fine."[46] It underlines that he is open to challenge or feedback, and that people won't be fired for disagreeing with him.

More important than any specific strength or weakness is your self-awareness. You need to know what your natural inclinations are in order to boost them or compensate for them.

THE ALIEN DIAGNOSTIC

If you want to challenge existing norms or conventions, you must first challenge your own assumptions. You must become more mindful of your habitual ways of thinking and behaving, more willing to believe you might be wrong.

Personality assessments and 360-degree feedback can expand your self-awareness, but they are often difficult to map against the five ALIEN thinking dimensions. To gauge your capacity to innovate, we've designed a behavior-based diagnostic test (see Figure 10.1 below). The test lets you rate your ALIEN thinking capacities. To assess the robustness of your creative process, fill out the questionnaire and calculate your average score for each element. Focus your attention on the lowest one or two scores.

The objective is not to undergo a personality change. It's to be yourself with more skill.[47] You need to be aware of your outlier tendencies and learn how they are perceived by others. The successful change makers in this book all had to work on themselves to achieve their aims.

Ultimately, the challenge is to apply ALIEN thinking to yourself and your development by paying attention to the current reality (your strengths and weaknesses, your habits, what you're doing, how you spend your time); levitating to reflect on what you want (what matters to you, what you believe, what gives you energy); imagining alternative trajectories (what else could you do, what options do you have, what constraints would you remove), experimenting with new ways of thinking or being (letting go of old habits and routines, trying new experiences that might help you evolve); and navigating change (anticipating obstacles, reaching out to people who can provide feedback or mentoring). You can turn to page 252 and take the ALIEN qualities quiz to assess your areas of strength and those requiring improvement or support from complementary partners.

To change the world, you must be willing to change yourself. You have an ALIEN thinking framework, but implementing it successfully starts with *you*.

⋅┤≻══┤══┤≺├⋅

KEY TAKEAWAYS

- Because of how the mind works, it's not easy to be an ALIEN thinker. In addition to the internal barriers to original thinking and creativity discussed earlier in the book, there is the challenge of managing your emotions and yourself.

- When setting off on a journey of innovation or discovery, you will have to overcome your fears about what *might* happen.

- You may also have to overcome "anticipatory regret"—imagining the regret you would feel if a decision or nondecision proves to be a mistake. One way is with "existential regret." Instead of envisioning failure, imagine how you would feel if you didn't try—if you didn't follow your heart because you decided to play it safe.

- Fear doesn't have to be a negative emotion or inhibiting force. Fear can also be a positive driver. Fear can push you to figure out the flaws in your idea. It can encourage you to look for people who can provide constructive criticism.

- Pinpointing the source of the fear is critical to addressing it. Only then can you start taking anxiety-reducing actions. Becoming an ALIEN thinker means altering your relationship to fear, learning to read it and channel it into action.

- The other big emotional derailer is frustration. It is the result of disappointments and regrets about what has happened or is not happening. To overcome frustration, step back from the situation and adopt a forensic mindset. Do a postmortem. What went wrong and why? What assumptions proved false? You may realize that the weaknesses overshadowed the strengths, but that doesn't invalidate the strengths.

- Besides falling afoul of your fears and negative emotions, you can also be undone by what you regard as your positive qualities. Beware of overplaying your strengths.

- Self-awareness is key to developing the coping strategies that will enable you to deliver breakthrough solutions without derailing.

- Personality assessments and 360-degree feedback can expand your self-awareness, but they are often difficult to map against the five ALIEN dimensions. To gauge your capacity to innovate, use Figure 10.1 to rate your ALIEN thinking capacities.

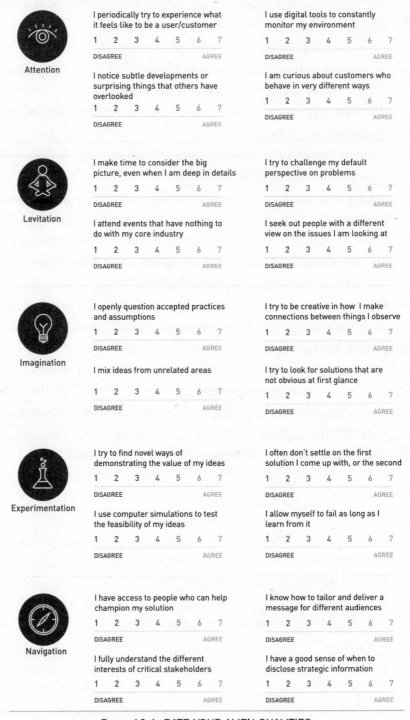

Attention

I periodically try to experience what it feels like to be a user/customer

1 2 3 4 5 6 7
DISAGREE   AGREE

I notice subtle developments or surprising things that others have overlooked

1 2 3 4 5 6 7
DISAGREE   AGREE

I use digital tools to constantly monitor my environment

1 2 3 4 5 6 7
DISAGREE   AGREE

I am curious about customers who behave in very different ways

1 2 3 4 5 6 7
DISAGREE   AGREE

Levitation

I make time to consider the big picture, even when I am deep in details

1 2 3 4 5 6 7
DISAGREE   AGREE

I attend events that have nothing to do with my core industry

1 2 3 4 5 6 7
DISAGREE   AGREE

I try to challenge my default perspective on problems

1 2 3 4 5 6 7
DISAGREE   AGREE

I seek out people with a different view on the issues I am looking at

1 2 3 4 5 6 7
DISAGREE   AGREE

Imagination

I openly question accepted practices and assumptions

1 2 3 4 5 6 7
DISAGREE   AGREE

I mix ideas from unrelated areas

1 2 3 4 5 6 7
DISAGREE   AGREE

I try to be creative in how I make connections between things I observe

1 2 3 4 5 6 7
DISAGREE   AGREE

I try to look for solutions that are not obvious at first glance

1 2 3 4 5 6 7
DISAGREE   AGREE

Experimentation

I try to find novel ways of demonstrating the value of my ideas

1 2 3 4 5 6 7
DISAGREE   AGREE

I use computer simulations to test the feasibility of my ideas

1 2 3 4 5 6 7
DISAGREE   AGREE

I often don't settle on the first solution I come up with, or the second

1 2 3 4 5 6 7
DISAGREE   AGREE

I allow myself to fail as long as I learn from it

1 2 3 4 5 6 7
DISAGREE   AGREE

Navigation

I have access to people who can help champion my solution

1 2 3 4 5 6 7
DISAGREE   AGREE

I fully understand the different interests of critical stakeholders

1 2 3 4 5 6 7
DISAGREE   AGREE

I know how to tailor and deliver a message for different audiences

1 2 3 4 5 6 7
DISAGREE   AGREE

I have a good sense of when to disclose strategic information

1 2 3 4 5 6 7
DISAGREE   AGREE

FIGURE 10.1: RATE YOUR ALIEN QUALITIES

QUESTIONS TO ASK YOURSELF

1. How good are you at managing your emotions?
2. What do other people (friends, family members, or colleagues) have to say about how well you manage emotions?
3. When was the last time you experienced anticipatory and/or existential regret, and why?
4. What is the worst thing that can happen if you fail—and how can you minimize/limit that risk?
5. Can you recall a time when you transformed fear into positive actions?

If you want to fight orthodoxies and improve your capacity to innovate, here's a good place to start. Fill out this questionnaire, then calculate your average score for each ALIEN dimension. Scores below 4 indicate areas for improvement. If there are several of these, focus your attention on the lowest one or two scores. To improve, draw inspiration from the individual chapters for novel perspectives or techniques. Equally, make sure you set aside time to consciously reflect on these activities periodically during the innovation journey. Ask yourself the six questions mentioned on pages 67–68. And look for complementary partners who can cover your weak spots.

Acknowledgments

WHILE THERE ARE THREE NAMES ON THE FRONT OF THIS BOOK, *ALIEN Thinking* is the result of a productive collaboration among a variety of talented people. Behind the author team is a crack unit of editors, graphic artists, marketers, media wizards, and administrators. Without the constant support of such amazing colleagues, *ALIEN Thinking* would never have been possible.

Many thanks to our developmental editor, Pete Gerardo, for his conscientiousness, clean writing, occasional criticism, and steady shepherding of the work. We're extremely grateful to our literary agent, Esmond Harmsworth, for his superb developmental advice, and for the great inspiration and positive energy he brought to us in important moments of the book-writing project. Our editor at Hachette/PublicAffairs, Colleen Lawrie, really helped to sharpen the manuscript with her keen eye for important details and surgical suggestions for improvement.

Marco Mancesti and Anand Narasimhan offered important and valuable oversight of the research activities at IMD. Successive IMD presidents, first Dominique Turpin and then Jean-François Manzoni, provided air cover where needed, as well as financial support for many of our activities. Our faculty colleagues—especially Howard Yu, Bill Fischer, Jim Pulcrano, and Theodore Peridis—provided invaluable insights, encouragement, and advice along the way. And outside of IMD, we are indebted to Estelle Metayer, Julian Birkinshaw, Alex Osterwalder, and Yves Pigneur; these truly original thinkers repeatedly informed and provoked our thinking, motivating us to explore a few interesting terrains through the writing journey.

We are extremely lucky to be able to continually test and refine our ideas with executives who come to IMD for professional development. Through this iterative approach, our work has been repeatedly challenged and improved. Therefore, we need to acknowledge and thank the thousands of executives who have been on the receiving end of our thought leadership. In many cases, they were catalysts and cocreators of this work, and their wisdom is at its heart.

Extensive writing and review sessions all take significant chunks of time, and thus we are grateful to our friends, colleagues, and especially our families for putting up with our frequent absences. Thank you all so very, very much.

Notes

INTRODUCTION

1. Brad Stone, *The Everything Store: Jeff Bezos and the Age of Amazon* (New York: Little, Brown, 2013), 234.

2. David Pierce, "The Kindle Changed the Book Business. Can It Change Books?," *Wired*, December 20, 2017.

3. Ron Adner, *The Wide Lens: What Successful Innovators See That Others Miss* (New York: Portfolio, 2012), 90–91.

4. Stone, *The Everything Store*, 238.

5. Adner, *The Wide Lens*, 96.

6. The concept of "vuja de" has been floating around the innovation field for a while, but the term was coined in 1984 by the edgy comedian George Carlin in one of his stand-up routines. He described it as "the strange feeling that, somehow, none of this has ever happened before." George Carlin, *George Carlin: Carlin on Campus*, directed by Steven J. Santos, filmed April 18–19, 1984, first aired on HBO.

7. Marcel Proust, *Remembrance of Things Past*, trans. C. K. Moncrieff, vol. 5 of 7, *The Captive* (London: Chatto and Windus, 1923), 69.

CHAPTER ONE: DISCOVERING THE DNA OF ORIGINALITY

1. Dr. Billy Fischer, presentation at the Orchestrating Winning Performance conference, Institute for Management Development, Lausanne, Switzerland, June 15, 2014.

2. Ibid.

3. Dr. Billy Fischer, interview by author, May 7, 2018.

4. Fischer, presentation.

5. Ibid.

6. Teresa Hodge, "Change Hiring Practices to Ensure Ex-felons Aren't Automatically Eliminated from the Running," *USA Today*, March 12, 2018.

7. Wendy Sawyer and Peter Wagner, "Mass Incarceration: The Whole Pie 2020" (press release), Prison Policy Initiative, March 24, 2020, https://www .prisonpolicy.org/reports/pie2020.html.

8. Teresa Hodge, "Teresa Hodge on Creating Opportunities for the Formerly Incarcerated," Racial Equity Video Series, SOCAP, June 4, 2020, www.social capitalmarkets.net/2020/06/racial-equity-video-series-teresa-hodge-on-creating -opportunities-for-the-formerly-incarcerated/.

9. Derek T. Dingle, "Success Beyond Bars: How Teresa Hodge Financially Empowers Formerly Incarcerated Entrepeneurs," *Black Enterprise*, September 21, 2018.

10. "About Us: A Note from Our Founder," R3 Score, last updated June 2020, https://www.r3score.com/f/About-Us/.

11. Teresa Hodge, "#61 Teresa Hodge, R3 and Mission:Launch," interview by Dave Dahl and Ladd Justesen, *Felony Inc. Podcast*, Startup Radio Network, June 5, 2019, https://soundcloud.com/felonyincpodcast/61-teresa-hodge -r3-and-missionlaunch.

12. "Mission: Launch Inc. Awarded 2015 SBA Growth Accelerator Prize to Unveil Business Accelerator for Formerly Incarcerated Persons," ImpactHub Washington DC, August 4, 2015, https://washington.impacthub.net/2015/08 /04/accelerator-missionlaunch/.

13. Teresa Hodge, "We Have Made Coming Home from Prison Entirely Too Hard," TEDx Talk, TEDxMidAtlantic conference "EVE: Everyone Values Equality," Washington, DC, November 16, 2015, YouTube video, 11:45, posted by TEDx Talks on April 8, 2016, https://www.youtube.com /watch?v=ibcgMS-0mAs.

14. Anne Field, "Startup's More-Nuanced Background Check Helps Ex-offenders Get Jobs," *Forbes*, July 17, 2019, https://www.forbes.com/sites /annefield/2019/07/17/startups-more-nuanced-background-check-helps-ex -offenders-get-jobs.

15. Ibid.

16. Ibid.

17. Hodge, "Change Hiring Practices."

18. Hodge, "We Have Made Coming Home."

19. Edie Weiner and Arnold Brown, *Future Think: How to Think Clearly in a Time of Change* (Upper Saddle River, NJ: Prentice Hall, 2006), 7–21.

20. Janet Landman, "Regret and Elation Following Action and Inaction: Affective Responses to Positive Versus Negative Outcomes," *Personality and Social Psychology Bulletin* 13, no. 4 (December 1987): 524–536.

21. Olga Craig, "James Dyson: The Vacuum Dreamer," *Telegraph*, August 24, 2008, https://www.telegraph.co.uk/finance/newsbysector/supportservices /2795244/James-Dyson-the-vacuum-dreamer.html.

22. Albert Meige and Jacques P. M. Schmitt, *Innovation Intelligence: Commoditization. Digitalization. Acceleration.* (n.p.: Absans Publishing, 2015).

23. For an extensive overview of the debate, see Cardiff Garcia, "Productivity and Innovation Stagnation, Past and Future: An Epic Compendium of Recent Views," *Financial Times*, March 11, 2016, http://ftalphaville.ft.com/2016/03 /11/2155269/productivity-and-innovation-stagnation-past-and-future-an-epic -compendium-of-recent-views/.

24. Gary Hamel, *What Matters Now: How to Win in a World of Relentless Change, Ferocious Competition, and Unstoppable Innovation* (San Francisco: Jossey-Bass, 2012).

25. CB Insights, *State of Innovation Report*, 2017, https://www.cbinsights .com/research-state-of-innovation-survey.

26. See the summary table of process models in R. Keith Sawyer, *Explaining Creativity: The Science of Human Innovation*, 2nd ed. (New York: Oxford University Press, 2012), 89.

27. Elmar Mock, "Reviving the Swiss Watch Industry: The Remarkable Story of Swatch," interview by Mark Bidwell, *OutsideVoices with Mark Bidwell* (podcast), December 19, 2016, https://innovationecosystem.libsyn.com/037-reviving -the-swiss-watch-industry-the-remarkable-story-of-swatch-with-elmar-mock.

28. Alexander Osterwalder and Yves Pigneur, *Business Model Generation: A Handbook for Visionaries, Game Changers, and Challengers* (Hoboken, NJ: John Wiley and Sons, 2010).

29. Jamie Williams, "Fischer Sees Medicine as a Way to Address Global Poverty," UNCGlobal, University of North Carolina at Chapel Hill, September 30, 2015, https://global.unc.edu/news-story/fischer-sees-medicine-as-a-way -to-address-global-poverty/.

30. Fischer, interview.

CHAPTER TWO: ATTENTION

1. Raymond Zhong, "Can an Edible Spoon Save the World?," *Wall Street Journal*, October 25, 2016.

2. Rakhi Chakraborty, "Eat What You Ate With: How Bakey's Is Combatting Plastic's War on the Environment with Edible Cutlery," YourStory.com, September 29, 2015, https://yourstory.com/2015/09/bakeys-edible-cutlery/.

3. Dibya Swetaparna Sarangi, "A Piquant Wit: Narayana Peesapaty," Monday Morning, National Institute of Technology, Rourkela, September 5, 2016, https://mondaymorning.nitrkl.ac.in/article/2016/09/05/782-a-piquant-wit-narayana -peesapaty/.

4. Zhong, "Can an Edible Spoon."

5. E. Byron, "More Pet Brands Target Owners Who Like to Cook Their Own Dog Food," *Wall Street Journal*, May 27, 2014.

6. Classroom discussion with Jouko Karvinen, Advanced Strategic Management program. IMD, Lausanne, Switzerland, November 5, 2014.

7. "Professor Yunus and the Origins of Grameen Bank," Grameen Italia Fondazione, n.d., http://www.grameenitalia.it/la-fondazione/professor-yunus -and-the-origins-of-grameen-bank/.

8. This quote and other details of the story are taken from Muhammad Yunus, *Creating a World Without Poverty: Social Business and the Future of Capitalism* (New York: PublicAffairs, 2007), 46.

9. Ibid.

10. Pam Henderson, "Can Innovation/Creativity Be Taught?," interview by David Robertson, *Innovation Navigation* (podcast), September 6, 2016, http://innovationnavigation.libsyn.com/9616-can-innovationcreativity-be -taught.

11. Yun Mi Antorini, Albert M. Muñiz Jr., and Tormod Askildsen, "Collaborating with Customer Communities: Lessons from the Lego Group," *MIT Sloan Management Review*, Spring 2012, 73–79.

12. Yun Mi Antorini and Albert M. Muñiz, "The Benefits and Challenges of Collaborating with User Communities," *Research-Technology Management* 56, no. 3 (May–June 2013): 21–28.

13. Simone Mitchell, "How IKEA Learned to Love IKEA Hacks (Almost as Much as We Do)," news.com.au, January 21, 2016, https://www.news.com.au /lifestyle/home/interiors/how-ikea-learned-to-love-ikea-hacks-almost-as-much -as-we-do/news-story/8f03b2779d6807315763ef56b0cd6fda.

14. John Wilbanks, "Unlocking Data and Unleashing Its Potential," panel on *Inside Social Innovation with SSIR* (podcast), June 5, 2017, https://ssir.org /podcasts/entry/unlocking_data_and_unleashing_its_potential.

15. Volker Bilgram, Michael Bartl, and Stefan S. Biel, "Getting Closer to the Consumer: How Nivea Co-creates New Products," *Marketing Review St. Gallen* 28, no. 1 (February 2011): 34–40.

16. Taylor Kubota, "Stanford Researchers Seek Citizen Scientists to Contribute to Worldwide Mosquito Tracking" (press release), *Stanford News Service*, October 31, 2017, https://news.stanford.edu/press-releases/2017/10/31/tracking -mosquitoes-cellphone/.

17. "The Top 8 Reddit Statistics on Users, Demographics & More," Mediakix, December 28, 2018, http://mediakix.com/2017/09/reddit-statistics-users-demographics/.

18. See the Adult Fans of LEGO subreddit at https://www.reddit.com/r/AFOL/.

19. Michael Wade, "Psychographics: The Behavioural Analysis That Helped Cambridge Analytica Know Voters' Minds," The Conversation, March 21, 2018, https://theconversation.com/psychographics-the-behavioural-analysis-that-helped-cambridge-analytica-know-voters-minds-93675.

CHAPTER THREE: LEVITATION

1. Neil Shea, "Swiss Adventurer Launches Quest to 'Fly Forever,'" *National Geographic*, March 6, 2015, https://www.nationalgeographic.com/news/2015/3/150306-solar-impulse-flight-pilot-circumnavigate-world-piccard-swiss/.

2. Bertrand Piccard, "My Ups and Downs with Solar Impulse" (blog post), July 24, 2016, https://bertrandpiccard.com/news/my-ups-and-downs-with-solar-impulse-539.

3. Ibid.

4. Shea, "Swiss Adventurer."

5. Carl Franklin, *Why Innovation Fails: Hard-Won Lessons for Business* (London: Spiro Press, 2003).

6. Frank Kalman, "Winter 2017 Insider . . . Kevin Kelly," *Talent Economy*, February 14, 2017, https://www.chieflearningofficer.com/2017/02/14/kevin-kelly/.

7. Mitsuru Kodama, "Managing Innovation Through Ma Thinking," *Systems Research and Behavioral Science* 35, no. 2 (March/April 2018): 155–177, https://doi.org/10.1002/sres.2453.

8. Randy L. Buckner, "The Serendipitous Discovery of the Brain's Default Network," *NeuroImage* 62, no. 2 (August 2012): 1137–1145.

9. Randy L. Buckner, "The Brain's Default Network: Origins and Implications for the Study of Psychosis," *Dialogues in Clinical Neuroscience* 15, no. 3 (September 2013): 351–358.

10. Marcus E. Raichle, "The Brain's Dark Energy," *Scientific American*, March 2010, 44–47.

11. Steven Johnson, *Where Good Ideas Come From: The Natural History of Innovation* (New York: Riverhead Books, 2010).

12. Kevin Davis, "Formerly Incarcerated People Are Building Their Own Businesses and Giving Others Second Chances," *ABA Journal*, July 1, 2019, https://www.abajournal.com/magazine/article/resolved-to-rebuild-formerly-incarcerated.

13. See Rosabeth Moss Kanter, "Innovation: The Classic Traps," *Harvard Business Review*, November 2006, 72–83.

14. Sally Maitlis and Marlys Christianson, "Sensemaking in Organizations: Taking Stock and Moving Forward," *Academy of Management Annals* 8, no. 1 (2014): 57–125.

15. Dean A. Shepherd, Jeffery S. McMullen, and William Ocasio, "Is That an Opportunity? An Attention Model of Top Managers' Opportunity Beliefs for Strategic Action," *Strategic Management Journal* 38, no. 3 (March 2017): 626–644.

16. Maitlis and Christianson, "Sensemaking in Organizations," 94.

17. Buckner, "The Serendipitous Discovery."

18. Darya Zabelina, Arielle Saporta, and Mark Beeman, "Flexible or Leaky Attention in Creative People? Distinct Patterns of Attention for Different Types of Creative Thinking," *Memory and Cognition* 44, no. 3 (2016): 488–498.

19. Shepherd, McMullen, and Ocasio, "Is That an Opportunity?"

20. William Ocasio, "Attention to Attention," *Organization Science* 22, no. 5 (September–October 2011): 1286–1296.

21. Shepherd, McMullen, and Ocasio, "Is That an Opportunity?"

22. Alison Beard and Sara Silver, "Life's Work: Ferran Adrià," *Harvard Business Review*, June 2011, 140.

23. Ibid.

24. Stefan Sagmeister, "The Power of Time Off," TED Talk, TEDGlobal 2009, July 2009, https://www.ted.com/talks/stefan_sagmeister_the_power_of _time_off.

25. Brian C. Gunia et al., "Contemplation and Conversation: Subtle Influences on Moral Decision Making," *Academy of Management Journal* 55, no. 1 (February 2012): 13–33.

26. Take, for example, the American singer Bebe Rexha, who won Best New Artist and Country Favorite Song awards at the 2018 Radio Disney Music Awards and has written numerous hits for others, including "The Monster" for Eminem and Rihanna and "Like a Champion" for Selena Gomez. Asked where she found inspiration for her songwriting, she said her favorite place was her bathroom: "A lot of those lyrics I wrote in the bathtub. I run the water and talk to myself and record it as a voice memo," she said. Kathy McCabe, "Meant to Be: Chart Slayer Bebe Rexha on Writing the Songs the Whole World Sings," news.com.au, July 11, 2018, https://www.news.com.au/entertainment/music /meant-to-be-chart-slayer-bebe-rexha-on-writing-the-songs-the-whole-world -sings/news-story/75ed9fa03feb7dcb00eaa80c79b48370. In another interview, she said that most of the songs on her award-winning album had emerged this way. "'I take a lot of baths,' says Rexha, laughing. She came up with most of *Expectations* in the tub, including 'Grace,' a piano-led song about a perfect other

who somehow doesn't fit the bill. 'I was sitting under the running water think-ing, "There's no easy way to break his heart. I could fly him to Paris and do it on top of the Eiffel Tower, but he's still gonna hate me no matter what."' She immediately grabbed her phone. 'It was soaking!'" Eve Barlow, "Anatomy of a Song: How Bebe Rexha Writes Hits," *Entertainment Weekly*, June 21, 2018, http://ew.com/music/2018/06/21/anatomy-of-a-song-bebe-rexha-expectations/.

27. Lisa Evans, "How Your 'Always-Busy' Pace Is Ruining Your Decision-Making," *Fast Company*, September 29, 2014, https://www.fastcompany.com/3036269/how-your-always-busy-pace-is-ruining-your-decision-making.

28. Kevin Cashman, "The Pause Principle," YouTube video, 8:21, August 15, 2014, https://www.youtube.com/watch?v=KVwj-mKDp5c.

29. Deniz Vatansever, David K. Menon, and Emmanuel A. Stamatakis, "De-fault Mode Contributions to Automated Information Processing," *Proceedings of the National Academy of Sciences of the United States of America* 114, no. 48 (November 2017): 12,821–12,826.

30. Roger E. Beaty et al., "Creativity and the Default Network: A Func-tional Connectivity Analysis of the Creative Brain at Rest," *Neuropsychologia* 64 (November 2014): 92–98.

31. Holly A. White and Priti Shah, "Scope of Semantic Activation and In-novative Thinking in College Students with ADHD," *Creativity Research Journal* 28, no. 3 (2016): 275–282.

32. Francine Kopun, "Adults with ADHD More Creative: Study," *Toronto Star*, February 9, 2011, https://www.thestar.com/life/health_wellness/diseases_cures/2011/02/09/adults_with_adhd_more_creative_study.html.

33. Charles Dickens to John Forster, 1854, in *The Letters of Charles Dick-ens*, Pilgrim Edition, ed. Madeline House, Graham Storey, and Kathleen Til-lotson, vol. 7, *1853–1855*, ed. Graham Storey, Kathleen Tillotson, and Angus Easson (New York: Oxford University Press, 1993), 429.

34. Flora Beeftink, Wendelien van Eerde, and Christel G. Rutte, "The Effects of Interruptions and Breaks on Insight and Impasses: Do You Need a Break Right Now?," *Creativity Research Journal* 20, no. 4 (2008): 358–364.

35. Daniel Pink, "Daniel Pink's 'When' Shows the Importance of Tim-ing Throughout Life," interview by Mary Louise Kelly, *All Things Considered*, NPR, January 17, 2018, https://www.npr.org/2018/01/17/578666036/daniel-pinks-when-shows-the-importance-of-timing-throughout-life.

36. K. Anders Ericsson, "The Influence of Experience and Deliberate Practice on the Development of Superior Expert Performance," chapter 38 in *The Cambridge Handbook of Expertise and Expert Performance*, eds. K. Anders Ericsson, Neil Charness, Robert R. Hoffman, and Paul J. Feltovich (New York: Cambridge University Press, 2006), 685–705.

37. Pink, interview.

38. Mary Helen Immordino-Yang, Joanna A. Christodoulou, and Vanessa Singh, "Rest Is Not Idleness: Implications of the Brain's Default Mode for Human Development and Education," *Perspectives on Psychological Science* 7, no. 4 (2012): 352–364.

39. Jonah Lehrer, "Creativity: Jonah Lehrer," interview by Andrew Marr, *Start the Week*, BBC Radio 4, April 30, 2012, https://www.bbc.co.uk/programmes /b01gnq8y.

40. Wilhelm Hofmann, Kathleen D. Vohs, and Roy F. Baumeister, "What People Desire, Feel Conflicted About, and Try to Resist in Everyday Life," *Psychological Science* 23, no. 6 (2012): 582–588, https://doi.org /10.1177/0956797612437426.

41. For large sample study results, see Roheeni Saxena, "The Social Media 'Echo Chamber' Is Real," Ars Technica, March 13, 2017, https://arstechnica .com/science/2017/03/the-social-media-echo-chamber-is-real/. For how echo chambers make you vulnerable to misinformation, see: Filippo Menczer, "Misinformation on Social Media: Can Technology Save Us?," The Conversation, November 27, 2016, https://theconversation.com/misinformation-on-social-media -can-technology-save-us-69264.

42. David Leonhardt, "You're Too Busy. You Need a 'Shultz Hour,'" *New York Times*, April 18, 2017, https://www.nytimes.com/2017/04/18/opinion/youre -too-busy-you-need-a-shultz-hour.html.

43. Adam Wernick, "One Woman's Plan to Take Your Creativity Back from Your Phone—by Making You Bored," *Studio 360*, PRI, January 31, 2015, https://www.pri.org/stories/2015-01-31/one-womans-plan-take-your-creativity -back-your-phone-making-you-bored.

44. Marlynn Wei, "What Mindfulness App Is Right for You?," *Huffington Post*, August 24, 2015, https://www.huffingtonpost.com/marlynn-wei-md-jd/what -mindfulness-app-is-right-for-you_b_8026010.html.

45. Emma Schootstra, Dirk Deichmann, and Evgenia Dolgova, "Can 10 Minutes of Meditation Make You More Creative?," *Harvard Business Review*, August 29, 2017, https://hbr.org/2017/08/can-10-minutes-of-meditation-make -you-more-creative.

46. "Digital Transformation at Axel Springer" (internal company video), Axel Springer, 2017.

47. Ibid.

48. Ibid.

49. Ibid.

50. Nicola Clark, "Axel Springer Reboots for Digital Age," *Irish Times*, January 4, 2016, https://www.irishtimes.com/business/axel-springer-reboots-for -digital-age-1.2475156.

CHAPTER FOUR: IMAGINATION

1. Warren Berger, *A More Beautiful Question: The Power of Inquiry to Spark Breakthrough Ideas* (New York: Bloomsbury USA, 2014), 34.

2. Ibid., 35.

3. Carol Pogash, "A Personal Call to a Prosthetic Invention," *New York Times*, July 2, 2008.

4. Ibid.

5. Online Etymology Dictionary, s.v. "imagine," accessed September 2, 2020, https://www.etymonline.com/word/imagine.

6. Drew Boyd, "Fixedness: A Barrier to Creative Output," *Psychology Today*, June 26, 2013.

7. Trevor MacKenzie with Rebecca Bathurst-Hunt, *Inquiry Mindset: Nurturing the Dreams, Wonders, and Curiosities of Our Youngest Learners* (n.p.: Elevate BooksEdu, 2018).

8. Ken Robinson, "Do Schools Kill Creativity?," TED Talk, TED2006 conference, Monterey, California, February 2006, https://www.ted.com/talks /ken_robinson_says_schools_kill_creativity.

9. Edward de Bono, *How to Have Creative Ideas* (London: Vermillion, 2008).

10. IDEO U, "Brainstorming—Rules & Techniques," n.d., https://www .ideou.com/pages/brainstorming-rules-and-techniques.

11. A large number of experimental studies have noted the disappointing creativity of brainstorming groups. One academic review noted: "In general, groups generate fewer and less creative ideas than do individuals working alone and judge relatively average ideas to be the most creative." Sarah Harvey and Chia-Yu Kou, "Collective Engagement in Creative Tasks: The Role of Evaluation in the Creative Process in Groups," *Administrative Science Quarterly* 58, no. 3 (2013): 346–386.

The failings of brainstorming are also reflected in the managerial press, where it is a recurrent theme and one that clearly resonates with practitioners. For example: Art Markman, "Your Team Is Brainstorming All Wrong," *Harvard Business Review*, May 18, 2017, https://hbr.org/2017/05/your-team-is-brainstorming -all-wrong; Tomas Chamorro-Premuzic, "Why Group Brainstorming Is a Waste of Time," *Harvard Business Review*, March 25, 2015, https://hbr.org/2015/03 /why-group-brainstorming-is-a-waste-of-time; Anne Fisher, "Why Most Brainstorming Sessions Fail," *Fortune*, August 23, 2013, http://fortune.com/2013 /08/23/why-most-brainstorming-sessions-fail/; and Natalie Peace, "Why Most Brainstorming Sessions Are Useless," *Forbes*, April 9, 2012, https://www .forbes.com/sites/nataliepeace/2012/04/09/why-most-brainstorming-sessions -are-useless.

Researchers who have tried to explain why interactive brainstorming frequently leads to mediocre results often point to "evaluation apprehension" as a key factor. This is a fear of negative evaluations from others that leads to self-censorship. Hassan Haddou, Guy Camilleri, and Pascale Zaraté, "Prediction of Ideas Number During a Brainstorming Session," *Group Decision and Negotiation* 23, no. 2 (2014): 271–298.

12. Jake Knapp, *Sprint: How to Solve Big Problems and Test New Ideas in Just Five Days* (New York: Simon and Schuster, 2016).

13. Donald G. McNeil Jr., "Car Mechanic Dreams Up a Tool to Ease Births," *New York Times*, November 13, 2013.

14. Ibid.

15. Patrick Bateson, "Playfulness and Creativity," *Current Biology* 25, no. 1 (January 5, 2015): R12–R16.

16. "Play and Creativity," hosted by Tom Sutcliffe, *Start the Week*, BBC Radio 4, February 13, 2017, https://www.bbc.co.uk/programmes/b08dmk4h.

17. Samira Far, "Why Your Playful Inner Child Is the Key to Innovation," *Inc.*, November 17, 2016.

18. Hal Gregersen, "Better Brainstorming," *Harvard Business Review*, March–April 2018, 64–71.

19. Berger, *A More Beautiful Question*; and Amanda Lang, *The Power of Why* (Toronto: HarperCollins, 2012).

20. Gonzalo Viña, "Big Data Promise Exponential Change in Healthcare," *Financial Times*, November 29, 2016.

21. "Meet Chris Sheldrick, Co-founder of What3Words," *Elle Decoration*, accessed July 21, 2018, http://elledecoration.co.za/chris-sheldrick/.

22. John Pollack, *Shortcut: How Analogies Reveal Connections, Spark Innovation, and Sell Our Greatest Ideas* (New York: Gotham Books, 2014). Also, in their book *Mental Leaps: Analogy in Creative Thought* (Cambridge, MA: MIT Press, 1999), cognitive scientists Keith Holyoak and Paul Thagard point out how many intellectual advances through the ages have been built on analogies.

23. Dedre Gentner, "Bootstrapping the Mind: Analogical Processes and Symbol Systems," *Cognitive Science* 34 (2010): 752–775.

24. Margalit Fox, "N. Joseph Woodland, Inventor of the Bar Code, Dies at 91," *New York Times*, December 12, 2012, https://www.nytimes.com/2012/12/13/business/n-joseph-woodland-inventor-of-the-bar-code-dies-at-91.html.

25. Kevin Dunbar, "How Scientists Think in the Real World: Implications for Science Education," *Journal of Applied Developmental Psychology* 21, no. 1 (2012): 49–58.

26. Jonathan Ledgard, conversation with author, Lausanne, Switzerland, June 19, 2014.

27. Nikolaus Franke, Marion K. Poetz, and Martin Schreier, "Integrating Problem Solvers from Analogous Markets in New Product Ideation," *Management Science* 60, no. 4 (2014): 1063–1081.

28. Lars Bo Jeppesen and Karim R. Lakhani, "Marginality and Problem-Solving Effectiveness in Broadcast Search," *Organization Science* 21, no. 5 (2010): 1016–1033.

29. Oguz Ali Acar and Jan van den Ende, "Knowledge Distance, Cognitive-Search Processes, and Creativity: The Making of Winning Solutions in Science Contests," *Psychological Science* 27, no. 5 (2016): 692–699.

30. Frans Johansson, "Innovating, Medici Style," interview by Mark Bidwell, *OutsideVoices with Mark Bidwell* (podcast), May 1, 2020, https://innovation ecosystem.libsyn.com/069-innovating-medici-style-with-frans-johansson.

31. Mike Butcher, "Potential New Treatment for COVID-19 Uncovered by BenevolentAI Enters Trials," *TechCrunch*, April 14, 2020, https:// techcrunch.com/2020/04/14/potential-new-treatment-for-covid-19-uncovered -by-benevolentai-enters-trials/.

32. Clive Cookson, "Biotechs Harness AI in Battle Against Covid-19," *Financial Times*, May 14, 2020, https://www.ft.com/content/877b8752-6847 -11ea-a6ac-9122541af204.

33. Paul Brackley, "BenevolentAI Founder Ken Mulvany on Using Artificial Intelligence to Find New Drugs to Tackle Disease," *Cambridge Independent*, March 10, 2018.

34. Michael Schrage, "Let Data Ask Questions, Not Just Answer Them," *Harvard Business Review*, October 8, 2014.

35. Wikipedia, s.v. "Analytical Engine," accessed September 2, 2020, https://en.wikipedia.org/wiki/Analytical_Engine.

36. Sean O'Neill, "How Creative Is Your Computer?," *New Scientist*, December 21, 2014.

37. For more details on these tests, see Jordan Pearson, "Forget Turing, the Lovelace Test Has a Better Shot at Spotting AI," *Vice*, July 8, 2014, https:// www.vice.com/en_us/article/pgaany/forget-turing-the-lovelace-test-has-a-better -shot-at-spotting-ai; and Selmer Bringsjord, Paul Bello, and David Ferrucci, "Creativity, the Turing Test, and the (Better) Lovelace Test," *Minds and Machines* 11, no. 3 (2001): 3–27, http://kryten.mm.rpi.edu/lovelace.pdf.

38. Demis Hassabis, "Exploring the Frontiers of Knowledge," DLD Conference, Munich, Germany, January 16, 2017, YouTube video, 26:27, posted by DLDconference on February 22, 2017, https://www.youtube.com/watch ?v=Ia3PywENxU8&t=962s.

39. Demis Hassabis, "What We Learned in Seoul with AlphaGo," *The Keyword* (blog), Google, March 16, 2016, https://blog.google/technology/ai /what-we-learned-in-seoul-with-alphago/.

40. James Vincent, "Google Uses DeepMind AI to Cut Data Center Energy Bills," The Verge, July 21, 2016, https://www.theverge.com/2016/7/21/12246258/google-deepmind-ai-data-center-cooling.

41. Lars Häggström, "Designing and Implementing Transformational Journeys: The Case of Stora Enso," presentation at Institute for Management Development, Lausanne, Switzerland, September 30, 2016.

42. Jouko Karvinen, "Looking to Past Values to Get Ahead of the Curve," interview by Paul Hunter, *Wednesday Webcast*, Institute for Management Development, January 13, 2013.

CHAPTER FIVE: EXPERIMENTATION

1. Jolene Creighton, "Business 101: Entrepreneurs Should Know No Barriers," Futurism.com, November 18, 2016, https://futurism.com/entrepreneurs-should-know-no-barriers.

2. Ibid.

3. Ibid.

4. Laurence Kemball-Cook, "Pavegen: How a Footstep's Energy Is Converted to Electrical Power," *The Edge*, CNBC, June 29, 2017, https://www.cnbc.com/video/2017/06/29/pavegen-how-a-footsteps-energy-is-converted-to-electrical-power.html.

5. Emma Gray Ellis, "The Best New Green Energy Tech Could Be Right Underfoot," *Wired*, June 13, 2016, https://www.wired.com/2016/06/best-new-gren-energy-tech-right-underfoot/.

6. Robert Nieman, "The Next Step in Renewable Energy Is Right Under Our Feet," Tech+, April 12, 2018, https://techplus.co/lehrer-changes-construction-practices-requires-holistic-approach-technology-2/.

7. Lydia Skrabania, "Pavegen: Generate Clean Electricity While Taking a Stroll," Reset, February 28, 2018, https://reset.org/node/29340.

8. Will Heilpern, "Meet the Clean-Tech CEO Who Compares His Company to Apple and Tesla," *Business Insider*, April 14, 2016, https://www.businessinsider.com/pavegen-kemball-cook-ceo-interview-2016-4.

9. CB Insights examined more than one hundred failed start-ups and discovered that the main culprit in most cases was that there wasn't a market need for the business idea.

10. Thomas R. Eisenmann, Eric Ries, and Sarah Dillard, "Hypothesis-Driven Entrepreneurship: The Lean Startup," HBS No. 812095-PDF-ENG (Cambridge, MA: Harvard Business Publishing, 2011).

11. Avinash Kaushik was the first to coin the term *HiPPO*, in his book *Web Analytics: An Hour a Day* (Indianapolis, IN: Wiley, 2007).

12. Albert Savoia, *Pretotype It: Make Sure You Are Building the Right "It" Before You Build "It" Right*, 2nd ed., October 11, 2011, http://www.pretotyping .org/uploads/1/4/0/9/14099067/pretotype_it_2nd_pretotype_edition-2.pdf.

13. Tarryn Leigh Lewis, Isabelle De Metz, and Lauranne Debbaudt, *Validation Guide: 24 Ways to Test Your Business Ideas*, n.d., Board of Innovation, http:// info.boardofinnovation.com/hubfs/Validation%20Guide%20compressed.pdf.

14. Ritch Macefield, "The Wizard of Oz Guide to Usability Testing Mobile Prototypes," *Userfocus*, May 1, 2012, https://www.userfocus.co.uk/articles/testing _mobile_prototypes.html.

15. Erno Tornikoski and Maija Renko, "Timely Creation of New Organizations: The Imprinting Effects of Entrepreneurs' Initial Founding Decisions," *M@n@gement* 17, no. 3 (2014): 193–213.

16. John Heilemann, "Reinventing the Wheel," *Time*, December 2, 2001.

17. Ibid.

18. Ibid.

19. Steve Kemper, *Code Name Ginger: The Story Behind Segway and Dean Kamen's Quest to Invent a New World* (Boston: Harvard Business School Press, 2003).

20. Ibid., 259.

21. Ibid., 273.

22. Madhumita Murgia, "Engineer Modifies Segway to Invent Hands-Free Wheelchair," *Telegraph*, October 20, 2015, https://www.telegraph.co.uk /technology/news/11942396/This-modified-Segway-is-a-hands-free-wheelchair -soon-to-be-on-sale.html.

23. Kemper, *Code Name Ginger*.

24. Bertrand Piccard, interview by Jean-François Manzoni at the Orchestrating Winning Performance conference, Institute for Management Development, Lausanne, Switzerland, June 29, 2018.

25. *The Simpsons*, season 16, episode 14, "The Seven-Beer Snitch," directed by Matthew Nastuk and David Silverman, written by Matt Groening, featuring Frank Gehry and Julie Kavner, aired April 3, 2005, on Fox in the US.

26. *Sketches of Frank Gehry*, directed by Sydney Pollack (Los Angeles: Sony Pictures Classics, August 22, 2006), DVD.

27. Ibid.

28. Paul Goldberger, *Building Art: The Life and Work of Frank Gehry* (New York: Alfred A. Knopf, 2015).

29. *Sketches of Frank Gehry*.

30. Richard J. Boland Jr., Fred Collopy, Kalle Lyytinen, and Youngjin Yoo, "Managing as Designing: Lessons for Organization Leaders from the Design Practice of Frank O. Gehry," *Design Issues* 24, no. 1 (2008): 10–25.

31. Jo Aggarwal, "An Interview with Jo Aggarwal, Co-inventor of Wysa," interview by Eric Wallach, *The Politic*, Yale University, March 28, 2018, http://thepolitic.org/an-interview-with-jo-aggarwal-co-inventor-of-wysa/.

32. Ramakant Vempati, "Show and Tell: Why the Product Matters Most," presentation at NASSCOM Product Conclave, Bangalore, India, November 2, 2017, YouTube video, 27:17, posted by NASSCOM Product on December 25, 2017, https://www.youtube.com/watch?v=sEVRqVEKmsQ.

33. Aggarwal, "An Interview with Jo Aggarwal."

34. Ibid.

35. Ibid.

36. Ed Catmull with Amy Wallace, *Creativity, Inc: Overcoming the Unseen Forces That Stand in the Way of True Inspiration* (London: Bantam Press, 2014), 92.

37. Ibid., 89–90.

38. Ibid., 99.

39. Ibid., 104.

40. Ibid., 94.

41. Ed Catmull, "How Pixar Fosters Collective Creativity," *Harvard Business Review*, September 2008, 64–72.

42. Catmull and Wallace, *Creativity, Inc.*,105.

43. Bernard Marr, "What Is Digital Twin Technology—and Why Is It So Important?," *Forbes*, March 6, 2017.

44. Patrick Wallis, "How NASA Mapped and Modeled Langley's Digital Twin," Esri, May 4, 2018, https://www.esri.com/about/newsroom/blog/nasa-langleys-digital-twin/.

45. Dan Perkel, "Digital Tools for Design Research," IDEO Labs, September 19, 2014, https://labs.ideo.com/2014/09/19/digital-tools-for-design-research/comment-page-2/.

46. Robin Johnson, "3 Real Life Examples of Incredibly Successful A/B Tests," HubSpot, February 6, 2018.

47. Tom Macaulay, "Real-World Use Cases for Google DeepMind's AI Systems," *ComputerWorld*, June 4, 2018.

48. David Morris, "Plaintiff: Theranos Used Shell Companies to Buy Outside Testing Equipment," *Fortune*, April 22, 2017, https://fortune.com/2017/04/22/theranos-shell-companies-suit/.

49. Darren Quick, "FTC Takes Action Against Crowdfunding Fraudster," New Atlas, June 15, 2015, https://newatlas.com/ftc-kickstarter-crowdfunding-fraud/38008/.

50. Nathan Furr and Jeffrey H. Dyer, "Leading Your Team into the Unknown," *Harvard Business Review*, December 2014, 80–88.

51. "SNCF: All Entrepreneurs" (internal company video), SNCF, 2016.

52. Ibid.

53. Ibid.

54. Lionel Steinmann, "Prix, organisation, qualité de service: les chantiers de Rachel Picard," *Les Echos*, June 27, 2016, https://www.lesechos.fr/2016/06 /prix-organisation-qualite-de-service-les-chantiers-de-rachel-picard-225400.

CHAPTER SIX: NAVIGATION

1. All quotes from Sarah Marquis are taken from her presentation at Institute for Management Development, Lausanne, Switzerland, June 28, 2019.

2. James Estrin, "Kodak's First Digital Moment," *New York Times*, August 12, 2015.

3. Steven Sasson, "Disruptive Innovation: The Story of the First Digital Camera," speech at Linda Hall Library, Kansas City, Missouri, October 26, 2011, Vimeo video, posted by Linda Hall Library on October 31, 2011, https:// vimeo.com/31404047.

4. Estrin, "Kodak's First Digital Moment."

5. Ibid.

6. Ibid.

7. Sasson, "Disruptive Innovation."

8. Ibid.

9. Ibid.

10. Douglas K. Smith and Robert C. Alexander, *Fumbling the Future: How Xerox Invented, Then Ignored, the First Personal Computer* (New York: William Morrow, 1988).

11. Sasson, "Disruptive Innovation."

12. Ibid.

13. Ibid.

14. Ibid.

15. Madison Malone-Kircher, "James Dyson on 5,126 Vacuums That Didn't Work—and the One That Finally Did," *New York*, November 22, 2016, http:// nymag.com/vindicated/2016/11/james-dyson-on-5-126-vacuums-that-didnt -work-and-1-that-did.html.

16. David Gram, "Becoming a Diplomatic Rebel," *IntraPRENEUR*, November 25, 2019, 76–81.

17. Annalisa Gigante, "Innovating a 2000 Year Old Product," interview by Mark Bidwell, *OutsideVoices with Mark Bidwell* (podcast), June 2, 2020, https://innovationecosystem.libsyn.com/049-innovating-a-2000-year-old -product-with-annalisa-gigante

18. Cyril Bouquet, "La poste se réinvente: Chronique d'une mutation stratégique réussie," *Harvard Business Review France*, September 9, 2014, https://www.hbrfrance.fr/chroniques-experts/2014/09/3270-la-poste-se-reinvente-chronique-dune-mutation-strategique-reussie/.

19. David Lagesse, "If Drones Make You Nervous, Think of Them as Flying Donkeys," *Goats and Soda* (blog), NPR, March 31, 2015, http://www.npr.org/sections/goatsandsoda/2015/03/31/395316686/if-drones-make-you-nervous-think-of-them-as-flying-donkeys.

20. Jake Colvin and Jordan Monroe, presentation at the 2013 International Business Model Competition, Harvard Innovation Lab, Boston, May 3, 2013, "IBMC 2013: Owlet – 1st Place," YouTube video, 17:55, posted by Business Model Competition Global on July 18, 2016, https://www.youtube.com/watch?v=f-8v_RgwGe0.

21. Tom Szaky, *Revolution in a Bottle: How TerraCycle Is Redefining Green Business* (New York: Portfolio, 2013), 94.

22. Ibid., 95.

23. Tom Szaky, "The Roller Coaster Ride of Entrepreneurship," presentation to students at Pennsylvania State University, April 18, 2013, YouTube video, 48:48, posted by BizStarts Milwaukee on August 27, 2013, https://www.youtube.com/watch?v=O5illbIh5m4.

24. Szaky, *Revolution in a Bottle*, 99.

25. Tom Szaky, "TerraCycle Founder on Why Purpose Isn't Enough for Social Entrepreneurship," interview with Andrew Warner, *Mixergy* (podcast), April 18, 2018, https://mixergy.com/interviews/TerraCycle-with-tom-szaky/.

26. Kim Bhasin, "The Incredible Story of How TerraCycle CEO Tom Szaky Became a Garbage Mogul," Business Insider, August 29, 2011, https://www.businessinsider.com/exclusive-tom-szaky-terracycle-interview-2011-8.

27. Tom Szaky, "Interview: Tom Szaky with TerraCycle," interview with Stone Payton and Lee Kantor, *High Velocity Radio* (podcast), March 20, 2018, https://businessradiox.com/podcast/highvelocityradio/terracycle/.

28. Szaky, "Purpose Isn't Enough."

29. Tom Szaky, "Revolutionizing Recycling One Cigarette Butt at a Time with TerraCycle's Tom Szaky," interview with Marjorie Alexander, *A Sustainable Mind* (podcast), November 30, 2017, https://asustainablemind.com/027-revolutionizing-recycling-one-cigarette-butt-at-a-time-with-terracycles-tom-szaky-2/.

30. Szaky, "Purpose Isn't Enough."

31. Ibid.

32. Bertrand Piccard, interview by Jean-François Manzoni at the Orchestrating Winning Performance conference, Institute for Management Development, Lausanne, Switzerland, June 29, 2018.

33. Ibid.

34. Navi Radjou and Jaideep Prabhu, *Frugal Innovation: How to Do Better with Less* (London: Profile Books, 2015), 57.

35. Szaky, "Purpose Isn't Enough."

36. Szaky, "Revolutionizing Recycling."

37. Tom Szaky, "Terracycle—Recycling 1 Million Pounds of 'Hard to Recycle' Materials Each Week," interview by Richard Jacobs, *Future Tech* (podcast), May 17, 2017, https://www.findinggeniuspodcast.com/podcasts /terracycle-recycling-1-million-pounds-of-hard-to-recycle-materials-each-week/.

38. Piccard, interview.

39. Ibid.

40. Jude Webber, "Lunch with the FT: Xavier López Ancona," *Financial Times*, August 1, 2014.

41. James L. Heskett, Javier Reynoso, and Karla Cabrera, "KidZania: Shaping a Strategic Service Vision for the Future," HBS No. 916402-PDF-ENG (Cambridge, MA: Harvard Business Publishing, 2015).

42. See "Franchises," KidZania, accessed September 3, 2020, http://www .kidzania.com/en/franchises.

43. "KidZania, the Success Game," May 29, 2015, https://www.mexico.mx /en/articles/kidzania-the-success-game.

44. Chris Sheldrick, "Meet the World Mapper Giving Everyone a New Address," interview by Anita Riotta, *Business Life* (podcast), Vox Markets, November 23, 2018, https://www.voxmarkets.co.uk/articles/chris-sheldrick -business-lives-fdb44ff/.

45. Viridiana Mendoza Escamilla, "KidZania está lista para conquistar (por fin) Estados Unidos," *Forbes México*, April 9, 2018, https://www.forbes.com .mx/kidzania-esta-lista-para-conquistar-por-fin-estados-unidos/.

46. Webber, "Xavier López Ancona."

47. David Güemes Castorena and José Alda Díaz Prado, "A Mexican Edutainment Business Model: KidZania," *Emerald Emerging Markets Case Studies* 3, no. 5 (2013), https://doi.org/10.1108/EEMCS-10-2013-0192.

48. Ibid.

49. Webber, "Xavier López Ancona."

50. Alistair Hall and Katie Ellman, "TerraCycle CEO Tom Szaky Makes Garbage the Hero," GreenBiz, November 18, 2016, https://www.greenbiz.com /article/terracycle-ceo-tom-szaky-makes-garbage-hero.

51. Bhasin, "The Incredible Story."

52. Szaky, *Revolution in a Bottle*, xviii.

53. Rebecca Mead, "When I Grow Up," *New Yorker*, January 19, 2015, https://www.newyorker.com/magazine/2015/01/19/grow.

54. Adam Minter, "Why Uber Is Losing Out to Locals in Southeast Asia," Livemint, July 27, 2017, https://www.livemint.com/Companies/ezkn3

YrxZiKH6Qs4W80lFN/Why-Uber-is-losing-out-to-locals-in-Southeast-Asia .html.

55. See "8-Step Process," John P. Kotter, n.d., https://www.kotterinc.com /8-steps-process-for-leading-change/.

56. Shruti Narula, "Case Study on Successful Viral Marketing of Old Spice," *Digital Marketing and Data Analytics Blog*, Digital Vidya, June 30, 2016, https:// www.digitalvidya.com/blog/case-study-on-successful-viral-marketing-of -oldspice/.

57. Associated Press, "Startups Shook Up the Sleepy Razor Market," CNBC, September 26, 2018, https://www.cnbc.com/2018/09/26/startups-shook -up-the-sleepy-razor-market-whats-next.html.

58. Ilyse Liffreing, "Vera Bradley Changes Course After Ads Are Labeled a 'Sexist Fail,'" Campaignlive.com, October 6, 2016, https://www.campaignlive .com/article/vera-bradley-changes-course-ads-labeled-sexist-fail/1411269.

59. Jennifer Jordan and Michael Sorell, "Why Reverse Mentoring Works and How to Do It Right," *Harvard Business Review*, October 3, 2019, https:// hbr.org/2019/10/why-reverse-mentoring-works-and-how-to-do-it-right.

60. All quotes from Darrell are taken from a conversation with the authors, June 26, 2019.

61. Jean-Philippe Deschamps and Michele Barnett Berg, "Logitech: Learning from Customers to Design a New Product," HBS No. IMD538-PDF-ENG (Lausanne, Switzerland: IMD, 2005).

62. All quotes from Donne-Crock are taken from a conversation with the authors, June 18, 2019.

CHAPTER SEVEN: THE ALIEN MODEL IN ACTION

1. Marc Levinson, a historian and economist, provides a fascinating account of the evolution of the project in his book *The Box: How the Shipping Container Made the World Smaller and the Economy Bigger* (Princeton, NJ: Princeton University Press, 2006).

2. John Winsor with Oguz A. Acar, "The Creative Potential of (Some) Outsiders," *Forbes*, April 25, 2017, https://www.forbes.com/sites/johnwinsor/2017 /04/25/the-creative-potential-of-some-outsiders/#13b75180b33c.

3. M. Canty, "The Ship That Never Calls at Port," *Maersk Post*, September 2016, 24–26, https://www.maersk.com/press/publications.

4. "Innovating Innovation," Delft Design Stories, Faculty of Industrial Design Engineering, TU Delft, May 4, 2017, https://www.tudelft.nl/en/ide /research/discover-design/innovating-innovation/.

5. For example, the Maersk Tankers Hackathon took place in Copenhagen on August 23–25, 2017. See http://maersktankershackathon.ilab.dk/.

6. Anneli Bartholdy, "Key Ingredients in Corporate Innovation: A Fireside Chat with Anneli Bartholdy, Maersk," interview by Peter Torstensen, Accelerance Fireside Chat, n.d., YouTube video, 25:00, posted by Accelerance DK, January 12, 2017, https://www.youtube.com/watch?v=_AmOqWls7P4.

7. Canty, "The Ship That Never Calls at Port."

8. Kevin Cashman, *The Pause Principle: Step Back to Lead Forward* (San Francisco: Berrett-Koehler Publishers, 2012).

9. François Englert, "How to Become a Nobel Prize Winner," interview by Christian Du Brulle, *Horizon*, December 10, 2013, https://horizon-magazine .eu/article/how-become-nobel-prize-winner_en.html.

CHAPTER EIGHT: A FLEXIBLE SEQUENCE

1. R. Keith Sawyer, *Explaining Creativity: The Science of Human Innovation*, 2nd ed. (New York: Oxford University Press, 2012), 89.

2. Bruce Nussbaum, "Design Thinking Is a Failed Experiment. So What's Next?," *Fast Company*, April 5, 2011, https://www.fastcompany.com/1663558 /design-thinking-is-a-failed-experiment-so-whats-next; Martin Kupp, Jamie Anderson, and Jörg Reckhenrich, "Why Design Thinking in Business Needs a Rethink," *MIT Sloan Management Review*, September 12, 2017, 42–44.

3. Mark Kurlansky, *Birdseye: The Adventures of a Curious Man* (New York: Doubleday, 2012).

4. Donald G. McNeil, "Car Mechanic Dreams Up a Tool to Ease Births," *New York Times*, November 13, 2013, https://www.nytimes.com/2013/11/14 /health/new-tool-to-ease-difficult-births-a-plastic-bag.html.

5. Bertrand Piccard, keynote speech, Orchestrating Winning Performance conference, Institute for Management Development, Lausanne, Switzerland, June 29, 2018.

6. Ibid.

7. Rachel Nuwer, "New Class of Polymers Discovered by Accident," *Scientific American*, December 1, 2014, https://www.scientificamerican.com/article /new-class-of-polymers-discovered-by-accident/.

8. Brian Chesky, "Interview with Airbnb CEO Brian Chesky," interview by Leigh Gallagher, *Fortune*, Economic Club of New York, n.d., YouTube video, 20:31, posted by Fortune Magazine on March 14, 2017, https://www.youtube .com/watch?v=GFMeuSIhIYg.

9. Elmar Mock, "Reviving the Swiss Watch Industry: The Remarkable Story of Swatch," interview by Mark Bidwell, *OutsideVoices with Mark Bidwell*, podcast, December 19, 2016, https://innovationecosystem.libsyn.com/037 -reviving-the-swiss-watch-industry-the-remarkable-story-of-swatch-with-elmar -mock.

10. Eliott C. McLaughlin, "Giant Rats Put Noses to Work on Africa's Land Mine Epidemic," CNN, September 8, 2010, http://edition.cnn.com/2010/WORLD/africa/09/07/herorats.detect.landmines/index.html.

11. Bart Weetjens, "First Person: 'I Teach Rats to Locate Landmines,'" interview by Jeremy Taylor, *FT Magazine*, February 7, 2015, 8–9.

12. McLaughlin, "Giant Rats Put Noses to Work."

13. Sharon Smith, "They Can Sniff Out Landmines and Detect TB—Meet the Rat Pack," *Times* (London), September 2, 2017, https://www.thetimes.co.uk/article/they-can-sniff-out-landmines-and-detect-tb-meet-the-rat-pack-5qxqxbpps.

14. Daisy Carrington, "Hero Rats Sniff (and Snuff) Out Landmines and TB," CNN, September 26, 2014, http://www.cnn.com/2014/09/26/world/africa/hero-rats-sniff-out-landmines-and-tb/index.html.

15. Smith, "Meet the Rat Pack."

16. Bart Weetjens, "Rats That Sniff Out Landmines and TB," TEDx Talk, TEDxBratislava, July 5, 2013, YouTube video, 14:02, posted by TEDx Talks on August 18, 2013, https://www.youtube.com/watch?v=E6atIJ8RDzU.

17. Jay Caboz, "Africa's Notorious Pest Becomes a Furry Savior," *Forbes*, November 20, 2014, https://www.forbes.com/sites/forbesinternational/2014/11/20/africas-notorious-pest-becomes-a-furry-savior/#5cfb2cce25c4.

18. David Vinjamuri, "Bic for Her: What They Were Actually Thinking (As Told by a Man Who Worked on Tampons)," *Forbes*, August 30, 2012, https://www.forbes.com/sites/davidvinjamuri/2012/08/30/bic-for-her-what-they-were-actually-thinking-as-told-by-a-man-who-worked-on-tampons/#53fd4dc33ab8. Ellen DeGeneres also satirized the Bic for Her pens on her talk show: "Bic Pens for Women," *The Ellen DeGeneres Show*, season 10, episode 25, aired October 12, 2012, on NBC, YouTube video, 4:08, posted by TheEllenShow on October 12, 2012, https://www.youtube.com/watch?v=eCyw3prIWhc.

CHAPTER NINE: DIGITAL

1. Melinda Rolfs, speech at Giving Innovation Summit, Urban Institute, Washington, DC, March 23, 2017, https://www.urban.org/events/giving-innovation-summit.

2. Sophia Bennett, "Mastercard Center Report Uses Big Data to Increase Charitable Donations," Sustainable Business, Conscious Connection, January 8, 2017, https://www.consciousconnectionmagazine.com/2017/01/mastercard-big-data-charitable-donations/.

3. Rolfs, speech at Giving Innovation Summit.

4. Melinda Rolfs, "Unlocking Data and Unleashing Its Potential," panel at Data on Purpose / Do Good Data: From Possibilities to Responsibilities conference, Stanford University, February 7–8, 2017, Vimeo video, posted by Stanford PACS on March 4, 2017, https://vimeo.com/206682124.

5. "Nonprofits Struggle to Understand Trends in Individual Giving," Mastercard Center for Inclusive Growth, December 14, 2016, https://www.mastercard center.org/insights/nonprofits-struggle-understand-trends-individual-giving.

6. Teresa Hodge, "Setting the Stage: Moral and Economic Obligations to Restoring Rights and Opportunity," panel discussion at the 17th Annual State Criminal Justice Network Conference, Atlanta, Georgia, August 23–25, 2018, YouTube video, 1:25:35, posted by NACDLvideo on June 4, 2019, https://www .youtube.com/watch?v=yzmXt539AwQ.

7. Ben Gomes, "Search Experiments, Large and Small," *Official Blog*, Google, August 26, 2008, https://googleblog.blogspot.ch/2008/08/search -experiments-large-and-small.html.

8. Rick Maese, "Moneyball 2.0: Keeping Players Healthy," *Washington Post*, August 24, 2015, https://www.washingtonpost.com/sports/moneyball-20-keeping -players-healthy/2015/08/24/5011ac54-48e6-11e5-9f53-d1e3ddfd0cda_story.html.

9. Raphael Rollier, conversation with author, Lusanne, Switzerland, March 4, 2020.

10. Ibid.

11. Daniel Kahneman, *Thinking, Fast and Slow* (New York: Farrar, Straus and Giroux, 2011), 10.

12. Hayley Matthews, "Online Dating Statistics: Dating Stats from 2017," Date Mix, December 3, 2017, https://www.zoosk.com/date-mix/online-dating -advice/online-dating-statistics-dating-stats-2017/.

13. Lauren Davidson, "These 3 Simple Questions Can Predict If an OkCupid Date Will Succeed," Mic, March 14, 2014, https://mic.com/articles /85297/these-3-simple-questions-can-predict-if-an-okcupid-date-will-succeed.

14. Richard E. Heyman and Amy M. Smith Slep, "The Hazards of Predicting Divorce Without Crossvalidation," *Journal of Marriage and Family* 63, no. 2 (2001): 473–479.

15. Antoine Gara, "BlackRock's Edge: Why Technology Is Creating the Amazon of Wall Street," *Forbes*, December 19, 2017, https://www.forbes.com /sites/antoinegara/2017/12/19/blackrocks-edge-why-technology-is-creating-a-6 -trillion-amazon-of-wall-street/.

16. Rachael Levy, "The COO at BlackRock Explains Why the $5.7 Trillion Investment Giant Is a 'Growth Technology Company,'" *Business Insider*, October 3, 2017, https://www.businessinsider.com/blackrock-coo-rob-goldstein -interview-2017-9.

17. Raffaele Savi and Jeff Shen, *Constant Change, Consistent Alpha: The Innovation Challenge for Active Investors*, BlackRock, October 2015.

18. Dan Schawbel, "Stanley McChrystal: What the Army Can Teach You About Leadership," *Forbes*, July 13, 2015, https://www.forbes.com/sites /danschawbel/2015/07/13/stanley-mcchrystal-what-the-army-can-teach-you -about-leadership/.

19. Ibid.

CHAPTER TEN: GREET YOUR INNER ALIEN

1. Bertrand Piccard, interview by Jean-François Manzoni at the Orchestrating Winning Performance conference, Institute for Management Development, Lausanne, Switzerland, June 29, 2018.

2. Dr. Billy Fischer, presentation at the Orchestrating Winning Performance conference, Institute for Management Development, Lausanne, Switzerland, June 2016.

3. Ibid.

4. Olga Craig, "James Dyson: The Vacuum Dreamer," *Telegraph*, August 24, 2008, https://www.telegraph.co.uk/finance/newsbysector/supportservices /2795244/James-Dyson-the-vacuum-dreamer.html.

5. Irving L. Janis and Leon Mann, "Anticipatory Regret," chap. 9 in *Decision Making: A Psychological Analysis of Conflict, Choice, and Commitment* (New York: Free Press, 1977), 219–242.

6. Daniel T. Gilbert, Carey K. Morewedge, Jane L. Risen, and Timothy D. Wilson, "Looking Forward to Looking Backward: The Misprediction of Regret," *Psychological Science* 15, no. 5 (2004): 346–350, https://doi.org/10.1111 /j.0956-7976.2004.00681.x.

7. Classroom discussion with the authors, June 28, 2019.

8. Marijo Lucas, "Existential Regret: A Crossroads of Existential Anxiety and Existential Guilt," *Journal of Humanistic Psychology* 44, no. 1 (2004): 58–70.

9. See Rakhi Chakraborty, "Eat What You Ate With: How Bakey's Is Combatting Plastic's War on the Environment with Edible Cutlery," YourStory.com, September 29, 2015, https://yourstory.com/2015/09/bakeys-edible-cutlery/.

10. Ibid.

11. Chris Sheldrick, "What3words' Chris Sheldrick: 'Driver, Take Me to "Table Chair Spoon,"'" interview by Danny Fortson, *Danny in the Valley* (podcast), June 28, 2018, https://play.acast.com/s/dannyinthevalley/chrisshelrick.

12. Ibid.

13. Ibid.

14. Teresa Y. Hodge, "Teresa Y. Hodge: Federal Prison Couldn't Stop My Relevance," interview by Marlon Peterson, *Decarcerated* (podcast), Decem-

ber 15, 2017, https://decarceratedpodcast.libsyn.com/teresa-y-hodge-federal
-prison-couldnt-stop-my-relevance.

15. Nikita Singareddy, "Former Prisoners Rethink Criminal Justice Through Entrepreneurship and Civic Technology," TechCrunch, September 4, 2015, https://techcrunch.com/2015/09/04/former-prisoners-rethink-criminal-justice
-through-entrepreneurship-and-civic-technology/.

16. Daniela Saderi (@Neurosarda), "'I have to disclose I'm not a data scientist, I'm not a researcher, I don't write code, but I have enough experience in business and I have the first hand . . . ,'" Twitter, January 12, 2019, 9:59 a.m., https://twitter.com/Neurosarda/status/1084102526331428864.

17. Jorge Odón, "Del taller mecánico a la sala de partos," TEDx Talk, TEDxRíodelaPlata, Buenos Aires, Argentina, May 2012, YouTube video, 11:47, posted by TEDxYouth on August 18, 2012, https://www.youtube.com/watch
?v=N-D8nt2EHQU.

18. Richard Branson, "Sir Richard Branson's Advice for Entrepreneurs: Don't Be Afraid of Fear," *Forbes*, April 16, 2015, https://www.forbes
.com/sites/realspin/2015/04/16/sir-richard-bransons-advice-for-entrepreneurs
-dont-be-afraid-of-fear/.

19. Ibid.

20. James Hayton and Gabriella Cacciotti, "How Fear Helps (and Hurts) Entrepreneurs," *Harvard Business Review*, April 3, 2018, https://hbr.org/2018/04
/how-fear-helps-and-hurts-entrepreneurs.

21. Piccard, interview.

22. Stephen Covey, A. Roger Merrill, and Rebecca R. Merrill, *First Things First: To Live, to Love, to Learn, to Leave a Legacy* (New York: Simon and Schuster, 1994), 59.

23. Bart Weetjens, "First Person: 'I Teach Rats to Locate Landmines,'" interview by Jeremy Taylor, *FT Magazine*, February 7, 2015, 89.

24. Laurence Kemball-Cook, "How to Change the World and Build a £20M Company from Your Bedroom," interview by Max Pepe, *Rebelhead Entrepreneurs* (podcast), July 25, 2016, http://rebelhead.com/2016/07/how-to
-change-the-world-and-build-a-20m-company-from-your-bedroom/.

25. Ibid.

26. Tom Szaky, "Interview: Tom Szaky with TerraCycle," interview with Stone Payton and Lee Kantor, *High Velocity Radio* (podcast), March 20, 2018, https://businessradiox.com/podcast/highvelocityradio/terracycle/.

27. Warren Berger, *A More Beautiful Question: The Power of Inquiry to Spark Breakthrough Ideas* (New York: Bloomsbury USA, 2014), 123.

28. Martha Davidson, "*Innovative Lives*: Artificial Parts: Van Phillips," Lemelson Center for the Study of Invention and Innovation, March 9, 2005, http://invention.si.edu/innovative-lives-artificial-parts-van-phillips.

29. Ibid.

30. Ginka Toegel and Jean-Louis Barsoux, "How to Become a Better Leader," *MIT Sloan Management Review* 53, no. 3 (2012): 50–60.

31. Bart Weetjens, "Conscious Leadership in Challenging Times," YouTube video, 2:21, posted by ashokaaustria on February 1, 2017, https://www.youtube.com/watch?v=IDOkvwJHwbo&t=29s.

32. Ferran Adrià, "Ferran Adrià on Auditing the Creative Process," interview by Ester Martinez and Suparna Chawla Bhasin, People Matters, November 12, 2018, https://www.peoplemattersglobal.com/article/innovation/ferran-adria-on-auditing-the-creative-process-19795.

33. Ferran Adrià, "Ferran Adrià on Transforming El Bulli from a Restaurant into a Legacy," interview by Isabel Conde, Eater, October 13, 2017, https://www.eater.com/2017/10/13/16435980/ferran-adria-interview-el-bulli-1846-bulligrafia-lab-museum.

34. Tim Hayward, "ElBulli for All," *FT Magazine*, June 21, 2013, https://www.ft.com/content/ad5f60de-d879-11e2-b4a4-00144feab7de.

35. Nick Bilton, "Why Google Glass Broke," *New York Times*, February 4, 2015, https://www.nytimes.com/2015/02/05/style/why-google-glass-broke.html.

36. Ibid.

37. Demis Hassabis, "DeepMind's Demis Hassabis," interview by Kamal Ahmed and Rohan Silva, *The Disrupters* (podcast), BBC Radio 4, November 6, 2018, https://www.bbc.co.uk/programmes/p06qvj98.

38. Ibid.

39. James Dyson, "Dyson: James Dyson," interview by Guy Raz, *How I Built This with Guy Raz* (podcast), NPR, February 12, 2018, https://www.npr.org/2018/03/26/584331881/dyson-james-dyson.

40. Hassabis, "DeepMind's Demis Hassabis."

41. John Carreyrou, *Bad Blood: Secrets and Lies in a Silicon Valley Startup* (New York: Alfred A. Knopf, 2018).

42. Hassabis, "DeepMind's Demis Hassabis."

43. Kevin Cashman, "Exploring the Power of Pause with Leadership Thought-Leader and Korn Ferry Senior Partner, Kevin Cashman," interview with Mark Bidwell, *OutsideVoices with Mark Bidwell*, April 6, 2020, http://innovationecosystem.com/power-of-pause-leadership-thought-leader-kevin-cashman/.

44. Tom Szaky, "TerraCycle Founder on Why Purpose Isn't Enough for Social Entrepreneurship," interview with Andrew Warner, *Mixergy* (podcast), April 18, 2018, https://mixergy.com/interviews/TerraCycle-with-tom-szaky/.

45. Ibid.

46. Ibid.

47. Robert Goffee and Gareth Jones, "Why Should Anyone Be Led by You?," *Harvard Business Review*, September–October 2000, 62–70.

Index